LONDON MATHEMATICAL SOCIETY LECTURE NOTE SERIES

Managing Editor: Professor I.M. James,
Mathematical Institute, 24-29 St Giles,Oxford

1. General cohomology theory and K-theory, P.HILTON
4. Algebraic topology, J.F.ADAMS
5. Commutative algebra, J.T.KNIGHT
8. Integration and harmonic analysis on compact groups, R.E.EDWARDS
9. Elliptic functions and elliptic curves, P.DU VAL
10. Numerical ranges II, F.F.BONSALL & J.DUNCAN
11. New developments in topology, G.SEGAL (ed.)
12. Symposium on complex analysis, Canterbury, 1973, J.CLUNIE & W.K.HAYMAN (eds.)
13. Combinatorics: Proceedings of the British Combinatorial Conference 1973, T.P.McDONOUGH & V.C.MAVRON (eds.)
15. An introduction to topological groups, P.J.HIGGINS
16. Topics in finite groups, T.M.GAGEN
17. Differential germs and catastrophes, Th.BROCKER & L.LANDER
18. A geometric approach to homology theory, S.BUONCRISTIANO, C.P. BOURKE & B.J.SANDERSON
20. Sheaf theory, B.R.TENNISON
21. Automatic continuity of linear operators, A.M.SINCLAIR
23. Parallelisms of complete designs, P.J.CAMERON
24. The topology of Stiefel manifolds, I.M.JAMES
25. Lie groups and compact groups, J.F.PRICE
26. Transformation groups: Proceedings of the conference in the University of Newcastle-upon-Tyne, August 1976, C.KOSNIOWSKI
27. Skew field constructions, P.M.COHN
28. Brownian motion, Hardy spaces and bounded mean oscillations, K.E.PETERSEN
29. Pontryagin duality and the structure of locally compact Abelian groups, S.A.MORRIS
30. Interaction models, N.L.BIGGS
31. Continuous crossed products and type III von Neumann algebras, A.VAN DAELE
32. Uniform algebras and Jensen measures, T.W.GAMELIN
33. Permutation groups and combinatorial structures, N.L.BIGGS & A.T.WHITE
34. Representation theory of Lie groups, M.F. ATIYAH et al.
35. Trace ideals and their applications, B.SIMON
36. Homological group theory, C.T.C.WALL (ed.)
37. Partially ordered rings and semi-algebraic geometry, G.W.BRUMFIEL
38. Surveys in combinatorics, B.BOLLOBAS (ed.)
39. Affine sets and affine groups, D.G.NORTHCOTT
40. Introduction to Hp spaces, P.J.KOOSIS
41. Theory and applications of Hopf bifurcation, B.D.HASSARD, N.D.KAZARINOFF & Y-H.WAN
42. Topics in the theory of group presentations, D.L.JOHNSON
43. Graphs, codes and designs, P.J.CAMERON & J.H.VAN LINT
44. Z/2-homotopy theory, M.C.CRABB
45. Recursion theory: its generalisations and applications, F.R.DRAKE & S.S.WAINER (eds.)
46. p-adic analysis: a short course on recent work, N.KOBLITZ
47. Coding the Universe, A.BELLER, R.JENSEN & P.WELCH
48. Low-dimensional topology, R.BROWN & T.L.THICKSTUN (eds.)

49. Finite geometries and designs, P.CAMERON, J.W.P.HIRSCHFELD & D.R.HUGHES (eds.)
50. Commutator calculus and groups of homotopy classes, H.J.BAUES
51. Synthetic differential geometry, A.KOCK
52. Combinatorics, H.N.V.TEMPERLEY (ed.)
53. Singularity theory, V.I.ARNOLD
54. Markov processes and related problems of analysis, E.B.DYNKIN
55. Ordered permutation groups, A.M.W.GLASS
56. Journées arithmétiques 1980, J.V.ARMITAGE (ed.)
57. Techniques of geometric topology, R.A.FENN
58. Singularities of smooth functions and maps, J.MARTINET
59. Applicable differential geometry, F.A.E.PIRANI & M.CRAMPIN
60. Integrable systems, S.P.NOVIKOV et al.
61. The core model, A.DODD
62. Economics for mathematicians, J.W.S.CASSELS
63. Continuous semigroups in Banach algebras, A.M.SINCLAIR
64. Basic concepts of enriched category theory, G.M.KELLY
65. Several complex variables and complex manifolds I, M.J.FIELD
66. Several complex variables and complex manifolds II, M.J.FIELD
67. Classification problems in ergodic theory, W.PARRY & S.TUNCEL
68. Complex algebraic surfaces, A.BEAUVILLE
69. Representation theory, I.M.GELFAND et. al.
70. Stochastic differential equations on manifolds, K.D.ELWORTHY
71. Groups - St Andrews 1981, C.M.CAMPBELL & E.F.ROBERTSON (eds.)
72. Commutative algebra: Durham 1981, R.Y.SHARP (ed.)
73. Riemann surfaces: a view toward several complex variables, A.T.HUCKLEBERRY
74. Symmetric designs: an algebraic approach, E.S.LANDER

London Mathematical Society Lecture Note Series: 74

Symmetric designs: an algebraic approach

ERIC S. LANDER
Assistant Professor, Harvard University

CAMBRIDGE UNIVERSITY PRESS
Cambridge
London New York New Rochelle
Melbourne Sydney

Published by the Press Syndicate of the University of Cambridge
The Pitt Building, Trumpington Street, Cambridge CB2 1RP
32 East 57th Street, New York, NY 10022, USA
296 Beaconsfield Parade, Middle Park, Melbourne 3206, Australia

First published 1983

Printed in Great Britain at the University Press, Cambridge

Library of Congress catalogue card number: 82-9705

British Library Cataloguing in Publication Data

Lander, Eric S.
Symmetric designs. - (London Mathematical Society
lecture notes series, ISSN 0076-0552; 74)
1. Symmetry groups
I. Title   II. Series
500   QD911

ISBN 0 521 28693 X

To my teachers, Peter J. Cameron
and John C. Moore
and to my wife, Lori Ann.

Δοσ πoυ στω ...

TABLE OF CONTENTS

PREFACE

Researchers studying the theory of error-correcting
des have discovered, in recent years, that finite geometries
d designs can provide the basis for excellent communications
hemes. The basic idea is to take the linear span (over some
nite field) of the rows of the incidence matrix of such a
ructure as the allowable messages. Mariner 9, for example,
ansmitted data to Earth by using a code derived from the
ructure of the hyperplanes in a five-dimensional vector space
er F_2, the field with two elements.

The purpose of this monograph is to allow coding
eory to repay some of its debt to the combinatorial theory
designs. Specifically, I have tried to show herein how the
jects introduced by coding theorists can offer great insight
to the study of symmetric designs.

The vector spaces and modules (over appropriate rings)
nerated by the incidence matrices of symmetric designs pro-
de a natural setting for invoking much algebraic machinery--
st notably, the theory of group representations--which has
therto not found much application in this combinatorial sub-
ct. In doing so, they provide a point of view which unifies
number of diverse results as well as makes possible many new
eorems. My own investigation into this subject is surely not
finitive, and if anyone is stimulated to further develop this
int of view, I will have accomplished something.

Two goals have informed my choice of organization.
rst, since my object is to expose a particular approach to
e study of symmetric designs, I have chosen to develop the
bject from scratch. This also seemed appropriate because it
ould make the topic accessible to readers who only know a

little combinatorics, and because no text yet exists squarely devoted to the subject of symmetric designs.

Second, it is my goal to advocate the increased use of powerful algebraic techniques in combinatorics. I realize that some of the combinatorialists who will read this monograph may not be familiar or comfortable with a number of the algebraic topics I shall employ. Rather than set prerequisites which might dissuade potential readers, I have included six appendices and two sections of supplementary problems which introduce and develop the techniques I shall require. These treatments are intended merely to be superficial and expedient --certainly not exhaustive. Rather, I hope the reader may choose to learn more about some of the topics by reading serious treatments of them later.

The only prerequisites are a first course in algebra (groups, rings, fields, modules); a smattering of number theory (through quadratic reciprocity, although an acquaintance with algebraic number theory would be helpful in a few places); and a knowledge of the basic counting principles and questions studied in combinatorics.

I have attempted to make this monograph useful to advanced undergraduates or graduate students as either a text or for self-study--as well as to the researcher active in the study of designs. Toward this end, I have included 126 problems to be solved, at the end of the chapters.

The text proceeds as follows. Chapter 1 contains an introduction to the basic notions about symmetric designs and provides a stock of examples upon which we draw throughout the text. Most of the material in this chapter is well-known to specialists, although a set of supplementary exercises developing an application of algebraic geometry to symmetric designs is new. Chapter 2 begins with the question of existence criteria for symmetric designs and proves the celebrated Bruck-Ryser-Chowla Theorem in the (more-or-less) standard way. Then we introduce the modules and vector spaces which are the main tools of this monograph, study their properties and use them to reinterpret the Bruck-Ryser-Chowla Theorem. Chapter studies automorphisms of symmetric designs, applying group representation theory to obtain theorems which supersede the previously known results due to Hughes. The methods are most effective in the case that a regular permutation group acts on the symmetric design in question and so in Chapter 4 we study the difference sets which arise in this case. Chapter 5 continues this study by presenting a number of multiplier theorems (including some new ones) for difference sets. An important aspect of Chapters 4 and 5 is that certain results previously thought to be unique to abelian difference sets are in fact special cases of more general results for symmetric designs. Finally, Chapter 6 raises some open questions concerning difference sets and presents tables of data concerning existence of particular difference sets. These tables are the product of my own hand calculations and I would be most grateful to anyone who can fill in any of the question marks that remain (or correct any errors I have made). The six appendices dealing with algebraic topics follow.

I warmly acknowledge my many debts. First and foremost, I owe much to Peter Cameron, my thesis supervisor, for all he has taught me--mathematically and otherwise--while advising me on the work which has led to this book; I hope he realizes how important his influence has been. I am also grateful to Peter Neumann for his many helpful conversations and for allowing me to wheedle him into teaching a wonderfully useful course on permutation modules in 1980 at Oxford.

I thank Jack van Lint for inviting me to the Technical University of Eindhoven, Netherlands, to give five weeks of lectures from an earlier version of this manuscript. I profited greatly from my contact with the members of his algebraic combinatorics seminar; in particular, I thank J. J. Seidel, H. Wilbrink, A. Brouwer, and A. Cohen. I also spent one of the most pleasant days of my mathematical career in Amsterdam with Hendrik Lenstra, who generously helped me sort out a number of questions. I also gratefully acknowledge the financial support of the Rhodes Trust, Wolfson College (Oxford), The Mathematical Institute (Oxford), and Harvard University.

Peter Cameron, Jack van Lint and Ed Assmus all provided very useful remarks on the manuscript. Susan Landau and Neil Immerman graciously agreed to assist in proofreading, for which I thank them. I, of course, am fully responsible for all errors--mathematical, calculational, and typographical --which remain.

To Pat Giersz, my secretary, who did an excellent job typing, retyping, and revising this book under trying circumstances, I offer my deep gratitude--although I suspect she would prefer a promise that I will not undertake another project like this for at least twelve months.

Finally, there is my wife, Lori. I do not know where to begin to thank her for the support, interest, and encouragement which made this project a reality and for her ability, willingness, and patience in discussing subject matter at length, despite having no training in mathematics. I count myself a lucky man, indeed.

1. SYMMETRIC DESIGNS

§1.1 DEFINITIONS AND SIMPLE EXAMPLES

An incidence structure consists simply of a set $\mathcal{P}$ of points and a set $\mathcal{B}$ of blocks, with a relation of incidence between points and blocks. Being of such a general nature, incidence structures arise naturally in all branches of mathematics. The particular sort which is the subject of this monograph--symmetric designs--arose first in the statistical theory of the design of experiments, but they rapidly have become objects of great combinatorial interest in their own right.

A symmetric (v,k,λ) design (or a symmetric design with parameters (v,k,λ)) is an incidence structure satisfying the following six requirements:

(1) There are v points.

(2) There are v blocks.

(3) Any block is incident with k points.

(4) Any point is incident with k blocks.

(5) Any two blocks are incident with λ points.

(6) Any two points are incident with λ blocks.

To exclude degenerate cases, we also insist that $k > \lambda$.

These axioms are not independent, and we shall explore them further in §1.4.

Example 1. The paradigmatic example of a symmetric design is represented in Figure 1.1. This symmetric design has parameters (7,3,1). The seven points are 1,2,3,4,5,6, and 7. The seven blocks are the sets {1,2,4}, {2,3,5}, {3,4,6}, {4,5,7}, {5,6,1}, {6,7,2} and {7,1,3}. A point p is incident with a block B if $p \in B$.

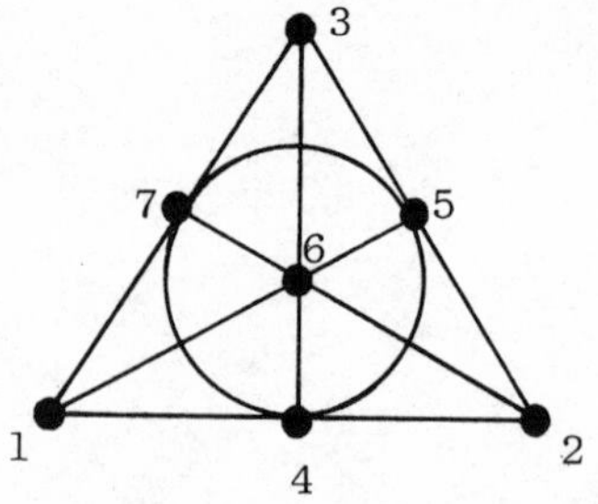

Figure 1.1 The Design Theorist's Coat of Arms

Example 2. Given any set $\mathcal{P}$ of v points, we can define a symmetric (v,v-1,v-2) design by letting $\mathcal{B}$ consist of the subsets of $\mathcal{P}$ having size v-1. A point p is defined to be incident with a block B if $p \in B$. Similarly, if $\mathcal{B}$ consists of the singleton subsets of $\mathcal{P}$ a symmetric (v,1,0) design results. Symmetric designs of these two sorts are called trivial.

The requirement that $k > \lambda$ implies that distinct blocks are incident with distinct point sets. So, we can always identify a block with the set of points with which it is incident and in general we shall not fuss over the formal distinction between them. Nevertheless, it is better not to define blocks as sets of points--for doing so would obliterate the formal duality between points and blocks (i.e., the fact that the axioms are unchanged if we reverse the role of points and blocks).

Example 3. Let $\mathcal{P}$ be the set of 11 residue classes modulo 11 and let B = {1,3,4,5,9} be the set of non-zero quadratic residues. Any two of the sets B, B+1, B+2,..., B+10 share two residue classes (where $B+i = \{x+i \mid x \in B\}$) and they may be taken as the 11 blocks of a symmetric (11,5,2) design.

Example 4. Let $\mathcal{P}$ be the set of 16 small squares in Figure 1.2. To each square there corresponds a block as in the figure--namely, the points incident with the block are other six squares in the same row or column. A symmetric ,6,2) design results.

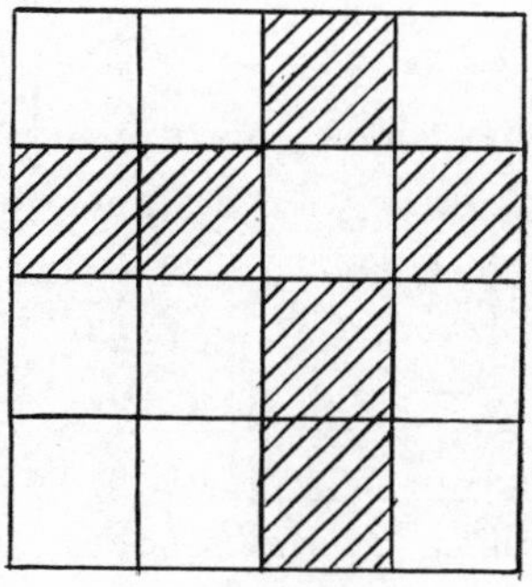

Figure 1.2

In this monograph we shall be concerned chiefly with loring for which triples (v,k,λ) there exist symmetric igns with parameters (v,k,λ) and exposing, along the way, ticularly interesting examples.

<u>Proposition 1.1</u>. <u>In a symmetric</u> (v,k,λ) <u>design</u>,

$$(1)\ (v-1)\lambda = k(k-1),$$

$$(2)\ k^2-v\lambda = k-\lambda, \text{ and}$$

$$(3)\ (v-k)\lambda = (k-1)(k-\lambda).$$

<u>Proof</u>. To prove the first assertion, choose a par- ular point q and count in two different ways the number of rs (p,B) with $p\neq q$ and B incident with both p and q (first summing over points and second by summing over blocks). second and third assertions are simply algebraic rearrange- ts of the first. Although they carry no new information, y will be useful for observing certain divisibility ations among the parameters.□

The value $k-\lambda$, which occurs in two of the equations ve, is an extremely important parameter. We set $n=k-\lambda$ call n the <u>order</u> of the symmetric (v,k,λ) design.

The <u>incidence matrix</u> A of a symmetric (v,k,λ) ign is the v x v matrix whose rows are indexed by blocks whose columns are indexed by points, with the entry in B and column p being 1 if p and B are incident and 0 erwise. (For our purposes, the particular order in which blocks and points are listed is irrelevant.) The incidence requirements can be expressed in terms of A:

$$AJ = JA = KJ \quad \text{and}$$

$$AA^T = A^TA = (k-\lambda)I + \lambda J = nI + \lambda J.$$

(Here, as throughout this book, I is the identity matrix and J the matrix with every entry 1, of appropriate size.) It is not difficult to show that the v x v matrix $aI + bJ$ has determinant $(a+vb)a^{v-1}$. Hence $\det(nI + \lambda J) = k^2 n^{v-1}$.

Proposition 1.2. *If A is the incidence matrix of a symmetric* (v,k,λ) *design then* $|\det A| = kn^{\frac{1}{2}(v-1)}$

Since det A must be an integer, we obtain our first substantial restriction upon the parameters of a symmetric design.

Theorem 1.3. (Schutzenberger [120]) *Suppose that there exists a symmetric* (v,k,λ) *design. If v is even, then n must be a square.*

An *isomorphism* from one symmetric design to another is a one-to-one mapping of points to points and blocks to blocks which preserves both incidence and nonincidence. Isomorphic symmetric designs necessarily have the same parameters. The condition is not sufficient however, and we shall soon see examples of symmetric designs with the same parameters which are not isomorphic.

Example 5. Reversing the roles of points and blocks in a symmetric (v,k,λ) design D, we obtain the *dual* of D, denoted D^{dual}, which is also a symmetric (v,k,λ) design. The incidence matrices of D and D^{dual} are transpose to one another. In general, D and D^{dual} need not be isomorphic.

Example 6. By taking the complement of the incidence relation in a symmetric (v,k,λ) design D (i.e., by replacing incidence by nonincidence and *vice versa*), we obtain the *complement* of D, denoted D'. The incidence structure D' is a symmetric (v', k', λ') design where $(v', k', \lambda') = (v, v-k, v-2k+\lambda)$. In particular, D and D' both have order $n=(k-\lambda)=(k'-\lambda')=n'$. A useful identity, yet another version of the fundamental equation $(v-1)\lambda=k(k-1)$, is the equation $\lambda\lambda'=n(n-1)$.

§1.2 HADAMARD MATRICES AND DESIGNS

A Hadamard matrix of order m is an m x m matrix H with entries ± 1 satisfying $HH^T = H^TH = mI$. (It is so named because its determinant attains a bound due to Hadamard.) These matrices are closely related to symmetric designs in a number of ways. Changing the sign of all entries in any row or column does not disturb the defining equation. We may therefore assume, if we like, that all entries in the first row and column are +1; call such a Hadamard matrix normalized. If we now delete the first row and column and replace -1 by 0 throughout we obtain a matrix M which (for $m \geq 4$) is the incidence matrix of a symmetric $(m-1, \frac{1}{2}m-1, \frac{1}{4}m-1)$ design. Such a symmetric design is called a Hadamard design.

$$\begin{pmatrix} +1 & +1 & +1 & +1 & +1 & +1 & +1 & +1 \\ +1 & -1 & +1 & -1 & +1 & -1 & +1 & -1 \\ +1 & +1 & -1 & -1 & +1 & +1 & -1 & -1 \\ +1 & -1 & -1 & +1 & +1 & -1 & -1 & +1 \\ +1 & +1 & +1 & +1 & -1 & -1 & -1 & -1 \\ +1 & -1 & +1 & -1 & -1 & +1 & -1 & +1 \\ +1 & +1 & -1 & -1 & -1 & -1 & +1 & +1 \\ +1 & -1 & -1 & +1 & -1 & +1 & +1 & -1 \end{pmatrix} \qquad \begin{pmatrix} 0&1&0&1&0&1&0 \\ 1&0&0&1&1&0&0 \\ 0&0&1&1&0&0&1 \\ 1&1&1&0&0&0&0 \\ 0&1&0&0&1&0&1 \\ 1&0&0&0&0&1&1 \\ 0&0&1&0&1&1&0 \end{pmatrix}$$

A normalized Hadamard matrix — Associated Hadamard design

Figure 1.3

From any symmetric design with parameters of the form $(m-1, \frac{1}{2}m-1, \frac{1}{4}m-1)$ we may in turn recover a normalized Hadamard matrix. (N.B. A Hadamard matrix may be modified by permuting rows and columns and subsequently renormalized; from such 'equivalent' Hadamard matrices nonisomorphic symmetric designs can result.)

The problem of constructing Hadamard matrices has received a great deal of attention and we shall only touch on a few basic methods. From the connection between Hadamard matrices and symmetric designs above, it follows almost immediately that a necessary condition for the existence of a Hadamard matrix of order m is that m=1, m=2 or m≡0 (mod 4). A longstanding conjecture states that this condition is sufficient as well. At present, no proof is known, but Hadamard matrices of order m have been found for all m divisible by 4 up through 264.

The unique normalized Hadamard matrix of order 2 is

$$H_2 = \begin{pmatrix} +1 & +1 \\ +1 & -1 \end{pmatrix}$$

If $H = (h_{ij})$ and K are Hadamard matrices of orders m and m', respectively, their Kronecker product

$$H \otimes K = \begin{pmatrix} h_{11}K & h_{12}K & \cdots & h_{1m}K \\ \vdots & & & \vdots \\ h_{m1}K & h_{m2}K & \cdots & h_{mm}K \end{pmatrix}$$

is a Hadamard matrix of order mm'. Starting with H_2, for example, we have the (normalized) Hadamard matrices

$$H_{2^r} = \underbrace{H_2 \otimes H_2 \otimes \cdots \otimes H_2}_{r \text{ times}}$$

which are called the Hadamard matrices of Sylvester type, or simply, the Sylvester matrices. (The normalized Hadamard matrix of Figure 1.3 is a Sylvester matrix.) Hence,

Example 7. There exist Hadamard designs with parameters $(2^m-1,\ 2^{m-1}-1,\ 2^{m-2}-1)$ for all integers $m \geq 2$.

There are two constructions which together allow us to associate to any finite field of odd characteristic a Hadamard matrix. Let q be a power of an odd prime p, say $q=p^f$. The Legendre symbol χ of $\mathbb{F}_q$ is the mapping $\chi: \mathbb{F}_q \longrightarrow \{0,1,-1\}$ given by

$\chi(0) = 0$

$\chi(a) = 1$ if a is a nonzero square in $\mathbb{F}_q$ and

$\chi(a) = -1$ if a is not a square in $\mathbb{F}_q$.

Note that $\chi(xy) = \chi(x)\chi(y)$. The Jacobsthal matrix $R = (r_{ij})$ is a q x q matrix whose rows and columns are indexed by the elements of $\mathbb{F}_q$ and in which $r_{ij} = \chi(i-j)$. The Jacobsthal matrix of order 7 is shown below.

$$\begin{pmatrix} 0 & 1 & 1 & -1 & 1 & -1 & -1 \\ -1 & 0 & 1 & 1 & -1 & 1 & -1 \\ -1 & -1 & 0 & 1 & 1 & -1 & 1 \\ 1 & -1 & -1 & 0 & 1 & 1 & -1 \\ -1 & 1 & -1 & -1 & 0 & 1 & 1 \\ 1 & -1 & 1 & -1 & -1 & 0 & 1 \\ 1 & 1 & -1 & 1 & -1 & -1 & 0 \end{pmatrix}$$

If $q \equiv 1 \pmod 4$ then -1 is a square in $\mathbb{F}_q$ and R is a symmetric matrix. If $q \equiv 3 \pmod 4$ then R is skew-symmetric; that is, $R^T = -R$.

Proposition 1.4. $R^TR = RR^T = qI - J$ and $RJ = JR = 0$.

Proof. The second equation reflects the fact that there are as many nonzero squares as nonsquares in $\mathbb{F}_q$. For a proof of the first, see Problems 6 and 7.□

Using the Jacobsthal matrices we can give constructions for Hadamard matrices, depending on the value of q (mod 4).

Example 8. Let q be a prime power congruent to 3 (mod 4). Let

$$H = \begin{pmatrix} 1 & 1 & \cdots & 1 \\ 1 & & & \\ \cdot & & & \\ \cdot & & & \\ \cdot & & (R - I) & \\ 1 & & & \end{pmatrix}.$$

Then

$$HH^T = \begin{pmatrix} q+1 & 0 & \cdots & 0 \\ 0 & & & \\ \cdot & & & \\ \cdot & & & \\ \cdot & & J+(R-I)(R^T-I) & \\ 0 & & & \end{pmatrix} = H^TH.$$

From Proposition 1.4 and the skew-symmetry of R, we find that H is a Hadamard matrix of order q+1, which is said to be of <u>Paley type</u>. Using this Hadamard matrix we obtain a Hadamard design with parameters $(q,\frac{1}{2}(q-1),\frac{1}{4}(q-3))$ which we denote H(q). (The incidence matrix of H(q) is of course obtained directly from R by replacing -1 by 0 throughout. That is, the blocks of H(q) are translates of the nonzero quadratic residues.)

H(11), which is Example 3 above, has incidence matrix

$$\begin{pmatrix} 0 & 1 & 0 & 1 & 1 & 1 & 0 & 0 & 0 & 1 & 0 \\ 0 & 0 & 1 & 0 & 1 & 1 & 1 & 0 & 0 & 0 & 1 \\ 1 & 0 & 0 & 1 & 0 & 1 & 1 & 1 & 0 & 0 & 0 \\ 0 & 1 & 0 & 0 & 1 & 0 & 1 & 1 & 1 & 0 & 0 \\ 0 & 0 & 1 & 0 & 0 & 1 & 0 & 1 & 1 & 1 & 0 \\ 0 & 0 & 0 & 1 & 0 & 0 & 1 & 0 & 1 & 1 & 1 \\ 1 & 0 & 0 & 0 & 1 & 0 & 0 & 1 & 0 & 1 & 1 \\ 1 & 1 & 0 & 0 & 0 & 1 & 0 & 0 & 1 & 0 & 1 \\ 1 & 1 & 1 & 0 & 0 & 0 & 1 & 0 & 0 & 1 & 0 \\ 0 & 1 & 1 & 1 & 0 & 0 & 0 & 1 & 0 & 0 & 1 \\ 1 & 0 & 1 & 1 & 1 & 0 & 0 & 0 & 1 & 0 & 0 \end{pmatrix}.$$

This symmetric design displays a great deal of 'symmetry.' We elaborate presently on this idea.

An <u>automorphism</u> of a symmetric design is an isomorphism of the symmetric design onto itself. To describe an automorphism then, we must give a permutation of the points and a permutation of the blocks which preserve the incidence structure. (So, in terms of the incidence matrix A, an automorphism specifies matrices P and Q such that PAQ = A.) The collection of all automorphisms forms a group under composition, called the <u>full automorphism group</u> of the symmetric design. Any subgroup is called an <u>automorphism group</u> of the symmetric design.

In practice, it is unnecessary to specify the action of an automorphism on both points and blocks. Once we know how it acts on points its action on blocks is determined--for blocks are completely specified by the points incident with them. Accordingly, we abuse terminology and frequently call a permutation π of the points of a symmetric design an automorphism if π induces an automorphism--that is, if for all blocks B, the set $\{\pi(p) \mid p \text{ is incident with } B\}$ is also a block.

Consider the Paley designs H(q). The permutation of the points given by $x \longmapsto x + b$, where b is any element of GF(q), induces an automorphism of H(q). The collection of all such automorphisms forms a group which acts regularly on the points (and blocks) of H(q). (The definitions and results which we shall require concerning permutation groups are summarized in Appendix A.) We can find an even larger automorphism group. Let $\Sigma(q)$ denote the group of all permutations π of the points given by

$$\pi: x \longrightarrow a\sigma(x) + b$$

where a is a nonzero square in F_q and σ is an automorphism of the field F_q. Since $\chi(\pi(i) - \pi(j)) = \chi(a)\chi(\sigma(i-j)) = \chi(i-j)$, every element of $\Sigma(q)$ induces an automorphism of H(q) and we may regard $\Sigma(q)$ as an automorphism group of H(q).

It is sometimes useful to work with the subgroup S(q) of $\Sigma(q)$ consisting of all the permutations π in which σ is the identity field automorphism. Note that S(q) has order $\frac{1}{2}q(q-1)$ and that when q is a prime, $S(q)=\Sigma(q)$.

The reader should check that both S(q) and $\Sigma(q)$ act 2-homogeneously (although not 2-transitively) and are thus rather large as permutation groups go.

Todd [130] first posed the question: Is $\Sigma(q)$ the full automorphism group of H(q)? Kantor [72] finally supplied the complete answer.

<u>Theorem 1.5</u>. (Kantor) <u>For</u> $q > 19$, <u>the full automorphism group of</u> H(q) <u>is</u> $\Sigma(q)$.

The proof goes beyond the scope of this book. For $q < 19$, it turns out that the full automorphism group of H(q) is even larger. When q=3, of course, the full symmetric group on three letters acts on H(3), which is a symmetric (3,2,1) design. In the other two cases, q=7 and q=11 we can find a 2-transitive automorphism group of $H(q)$ properly containing $\Sigma(q)$. We explore this in Problems 3 and 4.

Having now constructed Hadamard matrices and designs corresponding to F_q when $q \equiv 3 \pmod 4$, we turn to the case $q \equiv 1 \pmod 4$.

Example 9. Suppose that q is a prime power congruent to 1 (mod 4). Let

$$M = \begin{pmatrix} 0 & 1 & . & . & 1 \\ 1 & & & & \\ . & & & & \\ . & & R & & \\ . & & & & \\ 1 & & & & \end{pmatrix}$$

where R is the Jacobsthal matrix of F_q. Construct a Hadamard matrix of order $2(q+1)$ as follows. Define auxiliary matrices

$$U = \begin{pmatrix} 1 & 1 \\ 1 & -1 \end{pmatrix} \quad \text{and} \quad V = \begin{pmatrix} 1 & -1 \\ -1 & -1 \end{pmatrix}$$

Replace each 0 in M by V, each +1 in M by U and each -1 in M by -U. Using the relations

$$UU^T = VV^T = \begin{pmatrix} 2 & 0 \\ 0 & 2 \end{pmatrix}$$

$$\text{and} \quad UV^T = -VU^T = \begin{pmatrix} 0 & -2 \\ 2 & 0 \end{pmatrix}$$

we can verify that a Hadamard matrix results (Problem 8). From the Hadamard matrix, we obtain a symmetric $(2q+1, q, \frac{1}{2}(q-1))$ design which we denote by $H'(2q+1)$.

There is an entirely different construction of symmetric designs which relies not on normalized Hadamard matrices but on Hadamard matrices with constant row and column sums. Suppose that H is a Hadamard matrix of order v having constant row and column sums. We shall see that the positions of the +1 entries describe a symmetric design. For, let k be the number of +1 entries in each row (column). Select any pair of rows (columns) and let λ be the number of positions in which both contain +1. Because $HH^T = H^TH = vI$, we find that $v=4(k-\lambda)$. Hence λ is independent of the pair of rows (columns) chosen and we have a symmetric (v,k,λ) design, with $v=4n$. The construction is completely reversible. From any symmetric (v,k,λ) design with $v=4n$ we can recover a Hadamard matrix with constant row and column sums. Since v is even, n must be a square by Schutzenberger's Theorem. Say $n=N^2$. Reviewing the argument above, we find that $(v,k,\lambda) = (4N^2, 2N^2\pm N, N^2\pm N)$. Symmetric designs with $v=4n$ are called <u>H-designs</u> because of their connection with Hadamard matrices.

<u>Example 10</u>. The matrix

$$T = \begin{pmatrix} +1 & -1 & -1 & -1 \\ -1 & +1 & -1 & -1 \\ -1 & -1 & +1 & -1 \\ -1 & -1 & -1 & +1 \end{pmatrix}$$

is a Hadamard matrix with constant row and column sums. Taking Kronecker products preserves this property. Thus we obtain a sequence of symmetric $(2^{2t}, 2^{2t-1}\pm 2^{t-1}, 2^{2t-2}\pm 2^{t-1})$ designs for $t\geq 1$. These symmetric designs turn out to have rather remarkable properties (for one thing, the full automorphism groups act 2-transitively, as we shall see in Chapter 3) and we shall find them popping up from time to time in various guises. Example 4 is seen to be isomorphic to the design above with t=2.

§1.3 PROJECTIVE GEOMETRIES

A projective geometry of dimension m over a field F is, loosely speaking, the collection of subspaces of a vector space V of dimension m+1 over F. The <u>elements</u> of the geometry are as follows: the <u>points</u> are the 1-dimensional subspaces of V, the <u>lines</u> are the 2-dimensional subspaces,..., the <u>projective</u> j-<u>dimensional</u> <u>subspaces</u> are the (j+1)-dimensional subspaces,..., the <u>hyperplanes</u> are the m-dimensional subspaces. An element is said to be incident with any element containing it, or <u>vice</u> <u>versa</u>. (So, for example, any two points are incident with a unique line.)

When it is convenient to do so, we shall identify a projective j-dimensional subspace with the set of points incident with it. (We shall assume that the reader is somewhat acquainted with projective geometries and their associated groups. If not, skimming Biggs & White [15, pp. 24-45] is more than sufficient.)

Example 11. Let $F = F_q$. The points and hyperplanes of a projective geometry of dimension m over F form a symmetric design (when $m \geq 2$) with parameters $(q^m+...+q+1, q^{m-1}+...+q+1, q^{m-2}+...+q+1)$. The order of the symmetric design is q^{m-1}. We denote the design by PG(m,q).

A projective geometry of dimension 2 is called a projective plane. The parameters of PG(2,q) are $(q^2+q+1, q+1, 1)$. The symmetric (7,3,1) design of Example 1 is seen to be isomorphic to PG(2,2):

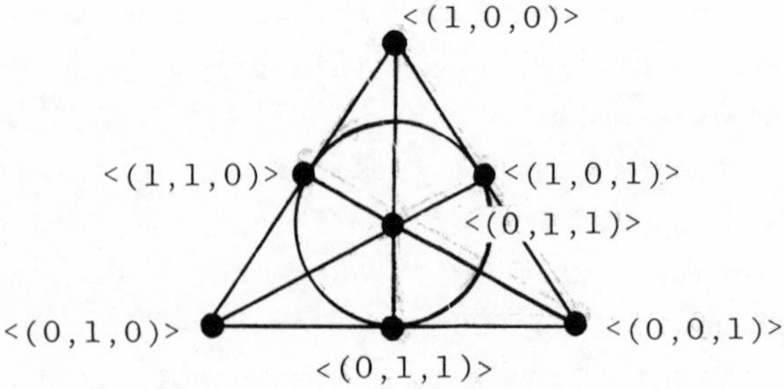

It is customary to broaden the term projective plane to include any symmetric design in which $\lambda=1$, even if it is not isomorphic to some PG(2,q). We shall adopt this custom.

The full automorphism group of PG(m,q) turns out to be quite interesting and important. Any permutation of the points which carries hyperplanes to hyperplanes must necessarily carry lines to lines--since the line through two points is the intersection of the hyperplanes containing them. The full automorphism group of PG(m,q) thus coincides with the full collineation group of the associated projective geometry (collineations are permutations of the points carrying lines to lines). The 'Fundamental Theorem of Projective Geometry' states that this group is induced by the group of semilinear mappings of the underlying vector space V (a *semilinear mapping* of V is a permutation ℓ of V such that for some automorphism α of the field F, we have

$$\ell(x+y) = \ell(x) + \ell(y) \quad \text{and}$$

$$\ell(uy) = \alpha(u)\ell(y)$$

for all $x, y \in V$ and $u \in F$). The full collineation group is called $P\Gamma L(m+1,q)$, the subgroup induced by linear mappings of V (i.e., those for which α is the identity field automorphism) is called $PGL(m+1,q)$, and the subgroup induced by linear maps whose determinant is unity is called $PSL(m+1,q)$.

We should remark that $PSL(m+1,q)$ acts as a 2-transitive permutation group on the points of the projective geometry and that it is a simple group.

Among all the projective-geometry point-hyperplane designs, my favourites are the PG(m,2). They are at once the most accessible and the most subtle. Throughout the text, they (or the associated vector space over F_2) shall continue to crop up. The parameters of PG(m,2) are $(2^{m+1}-1,\ 2^m-1,\ 2^{m-1}-1)$, which means that it is a Hadamard design. Is it isomorphic to any of the Hadamard designs we have already seen?

Proposition 1.6. *The Hadamard design obtained from the Sylvester matrix* $H_{2^{m+1}}$ *is isomorphic to* PG(m,2).

Proof. Let G_{2^m} be the matrix obtained by replacing -1 by 0 in H_{2^m}. We claim that the columns of G_{2^m} can be put into one-to-one correspondence with the elements $v_1,\ldots,v_{2^m}$ of the vector space $V_m = (F_2)^m$ in such a way that the rows of G_{2^m} are precisely the characteristic functions of the subspaces of dimensions m or m-1. (The *characteristic function* of a subset S of V_m is the vector which is 1 in columns corresponding to elements of S and 0 elsewhere.) For example, when m=2, we can do this as follows:

$$\begin{array}{cccc} (0,0) & (1,0) & (0,1) & (1,1) \\ \end{array}$$

$$\begin{pmatrix} 1 & 1 & 1 & 1 \\ 1 & 0 & 1 & 0 \\ 1 & 1 & 0 & 0 \\ 1 & 0 & 0 & 1 \end{pmatrix}$$

The first row is the characteristic function of the (unique) subspace of dimension 2 and the remaining rows are the characteristic functions of the subspaces of dimension 1. We proceed by induction on m. Assume that the columns of G_{2^m} correspond to the elements $v_1, \ldots, v_{2^m}$ of V_m in the desired way. Now, recalling the Kronecker product construction for $H_{2^{m+1}}$, we see that

$$G_{2^{m+1}} = \begin{pmatrix} G_{2^m} & G_{2^m} \\ G_{2^m} & J-G_{2^m} \end{pmatrix} .$$

Consider the vector space $V_{m+1} = V_m \oplus F_2$. Let the first 2^m columns of $G_{2^{m+1}}$ correspond to $(v_1,0), \ldots, (v_{2^m},0)$, respectively, and let the last 2^m columns correspond to $(v_1,1), \ldots, (v_{2^m},1)$. The subspaces of dimension m+1 or m in V_{m+1} are precisely the sets

$$(H,0) \cup (H,1)$$

$$\text{and} \qquad (H,0) \cup (V_m-H,1)$$

where H is a subspace of dimension m or m-1 in V_m (and for a subset, S in V we define $(S,i) = \{(s,i) \mid s \in S\}$).

By the inductive construction starting with G_4, the first column corresponds to the zero vector and the first row to the characteristic function of the entire vector space. Now, PG(m,2) is obtained merely by puncturing the zero vector of V_{m+1}. Hence, deleting the first row and column of $G_{2^{m+1}}$ yields the incidence matrix of points and hyperplanes (by way of the one-to-one correspondence). This proves the proposition.□

PG(m,2) also has the same parameters as $H(2^{m+1}-1)$, the latter being defined of course only when $2^{m+1}-1$ is a prime. For m=2, the symmetric designs are isomorphic (since there is only one symmetric design with 7 points; see Problem 3). For m>2, however, they are never isormorphic. One way to see this is to note that PG(m,2) has a 2-transitive automorphism group while (except for small q) the full automorphism group of H(q) is not 2-transitive, by Theorem 1.5. Unfortunately this argument rests on a result which we have quoted, but not proved. Accordingly, we offer a different, self-contained proof.

Consider, for example, PG(4,2) and H(31). We can show directly that they are not isomorphic. Choose two points x and y in PG(4,2). The intersection of all hyperplanes (i.e., blocks) containing them is precisely the unique line of the projective geometry passing through x and y; it has 3 points. On the other hand, a direct computation shows that the intersection of all blocks through any two points of H(31) contains exactly the two points. It turns out that this argument works in general: we can show that the intersection of all blocks through points x and y of H(q) is precisely {x,y}, except when q=7. While elementary, the proof is nontrivial; we include it as a set of supplementary problems to this chapter.

Inspired by this argument, we define the line through two points of a symmetric design to be the intersection of all blocks containing the two points. (Verify that this is a sensible definition by checking that any two distinct points are contained in a unique line.) The line through two points of PG(m,q) is precisely the line in the sense of projective geometry. Accordingly the lines of PG(m,q) behave quite nicely:

(1) every line has q+1 points.

(2) every line intersects every block.

(3) if x,y and z are three points not on a common line, then there are exactly $q^{m-3}+\dots+q+1$ blocks containing x, y and z.

In fact, each of these properties very nearly characterizes the point-hyperplane symmetric designs PG(m,q) in the sense of the following result.

Theorem 1.7. (Dembowski-Wagner [34]) Let D be a nontrivial symmetric (v,k,λ) design. The following are equivalent:

(1) Every line has exactly $h = (v-\lambda)/(k-\lambda)$ points.

(2) Every line has at least $h = (v-\lambda)/(k-\lambda)$ points.

(3) Every line meets every block.

(4) The number of blocks containing three noncollinear points is constant.

(5) D is either a projective plane or is isomorphic to some PG(m,q) (where $q=h-1$).

We shall prove this extremely important result in §1.5

The rich underlying geometric structure makes the symmetric designs PG(m,q) relatively easy to explore and to exploit. For example, suppose that we equip the underlying vector space with a nondegenerate symmetric bilinear form. (Bilinear forms are discussed in Appendix B.) The map σ: $U \longmapsto U^{\perp}$, sending each subspace to its perpendicular, is called polarity of the projective geometry. It exchanges points and blocks of PG(m,q) and preserves incidence and nonincidence. Thus, σ provides an isomorphism from PG(m,q) to its dual.

We close with two examples of symmetric designs which can be extracted from the geometry of PG(m,q).

Example 12. Let σ be a polarity (associated with a nondegenerate symmetric bilinear form) of a projective geometry of dimension 2m over F_q, where $m \geq 2$ and q is odd. Call a point p absolute if $p \in \sigma p$ and a hyperplane H is absolute if $\sigma H \in H$. The absolute points and hyperplanes of PG(2m,q) form a symmetric design, which we denote by A(2m,q). This symmetric design is a subdesign of PG(2m,q) in the sense that the points and blocks of the former are chosen from among those of the latter and the incidence relation is the restriction of incidence in PG(2m,q). The parameters of A(2m,q) are the same as those of PG(2m-1,q), but these designs are not isomorphic (see Problem 14).

Example 13. Further examples arise for q=3. Let Q be the quadratic form associated with σ in the previous example. Since Q(x) = Q(-x), the quadratic form Q induces a function from the projective points of PG(2m,q) onto F_3. If P_i is the set of projective points on which Q takes the value i (for $i=\pm 1$) and $B_i = \{\sigma(x) | x \in P_i\}$ then P_i and B_i are the points and blocks, respectively, of a symmetric design. The parameters are

$$(\tfrac{1}{2}3^m(3^m+i),\ \tfrac{1}{2}3^{m-1}(3^m-i),\ \tfrac{1}{2}3^{m-1}(3^{m-1}-i))$$

provided that the sign of Q is chosen correctly.

§1.4 t-DESIGNS

Symmetric designs are rather special examples of a broader class of incidence structure called t-designs. In this section we explore how symmetric designs relate to the more general theory and how they constitute an extreme case.

A t-(v,k,λ) design is an incidence structure satisfying the following requirements:

(1) There are v points.

(2) Any block is incident with k points.

(3) Any t points are incident with λ blocks.

To exclude trivialities, we insist that $k \neq v$ and $k \neq 0$.

Whenever $t \leq k$, we can find a rather uninteresting t-(v,k,λ) design in which the blocks are precisely the sets of cardinality k. We call this a complete t-design. All other t-designs are incomplete.

Unlike the situation for symmetric designs, distinct blocks may be incident with the same point set. If this occurs we say that the t-design has repeated blocks. For example, if we let each block of a t-design occur twice as often, the resulting structure is still a t-design, with twice as many blocks. Designs without repeated blocks are more interesting from a combinatorial point of view, although it is sometimes convenient to allow repetitions. (By contrast, statisticians do not mind repeated blocks in experimental design; fewer distinct experiments need be set up.)

Some examples of incomplete t-designs without repeated blocks are:

(1) A symmetric (v,k,λ) design is a 1-(v,k,k) design which is incomplete if $1<k<v-1$.

(2) A symmetric (v,k,λ) design is clearly a 2-(v,k,λ) design. Projective geometries supply many further examples: the points and projective j-dimensional subspaces of an m-dimensional projective geometry over any F_q (with $1<j<m$) form a 2-design which we denote $P_j(m,q)$.

Often 2-designs are called simply block designs or (if not complete) balanced incomplete block designs or BIBDS, in the literature.

(3) A 3-$(m,\frac{1}{2}m,\frac{1}{2}m-1)$ design may be constructed from any normalized Hadamard matrix H of order m. The columns represent the points. The positions of the +1's in each of the m-1 rows (apart from the first) give m-1 blocks. The position of the -1's give another m-1 blocks. Together these 2m-2 blocks form a 3-design (prove this) called a Hadamard 3-design.

(4) Some incomplete 4-designs and 5-designs without repeated blocks are known, but they are sufficiently rare that new constructions are interesting; none are known for $t\geq 6$. (If we allow repeated blocks, however, t-(v,k,λ) designs can be constructed for all t,v,k with $t<k<v-t$. See Problem 17.) We shall construct a 5-$(24,8,1)$ design in Chapter 2.

Proposition 1.8. A t-(v,k,λ) design is also an s-(v,k,λ_s) design for any $0\leq s\leq t$, where

$$\lambda_s = \lambda\frac{(v-s)(v-s-1)\ldots(v-t+1)}{(k-s)(k-s-1)\ldots(k-t+1)}$$

Proof. Choose a set of s points ($0\leq s\leq t$) and let the number of blocks incident with them be λ_s. Counting in two ways the number of choices of t-s further points and a block containing the set distinguished points, we obtain

$$\lambda_s \binom{k-s}{t-s} = \binom{v-s}{t-s}\lambda .$$

us λ_s is independent of the choice of the s points and is ual to the expression given in the theorem.□

The parameters λ_0 (the number of blocks) and λ_1 (the mber of blocks incident with a point) are usually denoted b and r, respectively. With t=1, s=0, the proposition ove shows that in any 1-design,

$$bk=vr.$$

any 2-design, the proposition implies that

$$r(k-1) = (v-1)\lambda .$$

n a symmetric design, v=b and r=k. Therefore we recognize is equation as a generalization of Proposition 1.1 (1).)

The incidence matrix B for a t-(v,k,λ) design is fined just as in the case of a symmetric design; B is a x v matrix. In a 2-design, the conditions that any block incident with k points, any point is incident with r ocks and any pair of points with λ blocks can be expressed terms of B:

$$BJ = kJ,$$

$$JB = rJ,$$

$$B^TB = (r-\lambda)I + \lambda J.$$

We compute $\det((r-\lambda)I + \lambda J) = rk(r-\lambda)^{v-1}$. Thus r>λ, then B^TB is nonsingular and B must have rank at least Since B has size b x v, we have :

<u>Proposition 1.9</u>. (Fisher's Inequality) <u>In a design</u>, $b \geq v$.

We mention without proof a generalization of sher's inequality due to Petrenjuk [111] in the case of t=2 d Ray-Chandhuri and Wilson [113] in general.

<u>Proposition 1.10</u>. <u>In a</u> 2t-<u>design</u>, $b \geq \binom{v}{t}$.

The <u>dual</u> of a t-design D is the incidence structure ual obtained by interchanging the roles of points and ocks. In general, the dual of a t-design is only a design. However a symmetric design is an example of a design whose dual is also a 2-design.

Theorem 1.11. In a 2-(v,k,λ) design, the following are equivalent:

(1) $b=v$,

(2) $r=k$,

(3) any two blocks meet in λ points,

(4) any two blocks meet in a constant number of points.

Proof. (1) implies (2): follows from $bk=vr$. (2) implies (3): suppose that (2) holds. Then $b=v$ and $BJ=JB$. Thus B commutes with $(r-\lambda)I + \lambda J$ and so with $((r-\lambda)I + \lambda J)B^{-1} = B^T$. Thus $BB^T = (r-\lambda)I + \lambda J$, from which we see that any two blocks are incident with λ points. (3) implies (4): Clear. (4) implies (1): if (4) holds, then D^{dual} is a 2-design. By Fisher's inequality, applied to D^{dual}, we have $b \leq v$. Thus $v = b$.□

Corollary 1.12. If D is a 2-design such that D^{dual} is also a 2-design, then D is a symmetric design.

Let D be a t-(v,k,λ) design, with any set of s points contained in λ_s blocks (for $s=0,1,\ldots,t$). The derived design D_p with respect to a point p is the $(t-1)$-design whose points are the points of D different from p and whose blocks are the sets $B-\{p\}$ where B is a block of D containing p. The parameters are $(t-1)$-$(v-1,k-1,\lambda)$. The residual design D^p with respect to p has the same point set but the blocks are those blocks of D not containing p. It is a $(t-1)$-$(v-1,k,\lambda_{t-1}-\lambda)$ design. For example, the derived design of a Hadamard 3-design is a Hadamard (symmetric) design; the residual design is the complement of the symmetric design.

A converse, which is important in the theory of permutation groups, is the question of extendability: given a t-design, is it isomorphic to D_p for some $(t+1)$-design? An extension may not exist, or there may be more than one. (To construct an extension, we must find a suitable design to play the role of D^p.) Every Hadamard (symmetric) design is (uniquely) extendable to a Hadamard 3-design. In the next chapter, we shall construct three different (but isomorphic) extensions of PG(2,4).

We can also construct derived and residual designs f D with respect to a block. However, in general, the esult is only a 0-design (that is, a ragtag collection of ubsets). An important exception is when D is a symmetric v,k,λ) design. Working dually, the *derived design* D_B with espect to a block B should be taken to be the incidence tructure whose points are the points of B and whose blocks re the sets B $\cap$ C, where C is any block other than B. The tructure is a 2-(k,λ,λ-1) design. The *residual design* D^B s the incidence structure whose points are those points of not on B and whose blocks are the sets C-B where C ranges ver all blocks other than B. It is a 2-(v-k,k-λ,λ) design. ither D_B or D^B may contain repeated blocks. (But see roblem 9.)

As an example, consider D = PG(m,q). The derived esign with respect to any block is the incidence structure nduced on any hyperplane. It looks exactly like PG(m-1,q), xcept that every block occurs q times. The residual design, n the other hand, has no repeated blocks. Its parameters re 2-(q^m,q^{m-1},q^{m-1}+...+q+1). Suppose that we have taken the esidual with respect to a hyperplane H. If J and K are two urther hyperplanes, then H$\cap$J$\cap$K is a projective subspace of imension (m-2) or (m-3). If m-2, then J-H and K-H are isjoint blocks in the residual design; if m-3, then J-H nd K-H meet in q^{m-2} points. Call two blocks *parallel* if hey do not meet. From the projective geometry we can check hat parallelism is an equivalence relation and that all quivalence classes have size q.

The points of the residual design can in fact be ut in one-to-one correspondence with the points of an m-imensional vector space over $\mathbb{F}_q$ in such a way that the locks are the translates of the (m-1)-dimensional subspaces. e call this structure an *affine geometry* and denote it G(m,q).

Motivated by this example, call a 2-(v,k,λ) design *ffine* if there exist integers s and μ such that the blocks an be partitioned into "parallel classes" of size s with the properties that

(1) blocks in the same parallel class are disjoint;

(2) blocks in different parallel classes meet in μ points.

A counting argument shows that $s=v/k$ and $\mu=v/s^2=k^2/v$. (See Problem 15.) Affine geometries are clearly affine designs. Hadamard 3-(v,k,λ) designs provide another example. By construction, the complement of a block is again a block, giving parallel classes of size 2. Non-parallel blocks share $v/4$ points.

The next theorem leads this digression about t-designs back to our main topic.

Theorem 1.13. *Suppose that there exists an affine design with parameters* $v,b,r,k,$ *and* λ. *Then there exists a symmetric* (v^*,k^*,λ^*) design with

$$v^* = (r+1)v,\ k^* = kr \quad \text{and} \quad \lambda^* = k\lambda.$$

Proof. Let Σ be the affine design. The blocks are partitioned into parallel classes of, say, s blocks each. Since $s=v/k$, there must be $bk/v=r$ parallel classes. We denote them by $\Pi_1,\ldots,\Pi_r$. Also, denote the points $p_1,\ldots,p_v$. To each parallel class Π_h we associate a matrix $M_h = (m^h_{ij})$ by the rule

$$m^h_{ij} = \begin{cases} 1 & \text{if } p_i \text{ and } p_j \text{ lie on some block in } \Pi_h \\ 0 & \text{otherwise} \end{cases}$$

We observe that

$$M_h = (M_h)^T,$$

$$M_h M_h^T = kM_h, \text{ and}$$

$$M_g M_h^T = \mu J, \quad \text{for } g \neq h.$$

(To see this, note that the entry in the i,j-th position of $M_g M_h^T$ is the number of points lying on both the unique block of Π_g through p_i and the unique block of Π_h through p_j.) Using the properties of Σ, we see that

$$\sum_{h=1}^{r} M_h = (r-\lambda)I + \lambda J,$$

nd thus

$$\sum_{h=1}^{r} (M_h M_h^T) = k(r-\lambda)I + k\lambda J.$$

ow form the (r+1)v x (r+1)v matrix

$$L = \begin{pmatrix} 0 & M_1 & M_2 & \dots & M_r \\ M_r & 0 & M_1 & \dots & M_{r-1} \\ \vdots & & & & \vdots \\ M_1 & M_2 & M_3 & \dots & 0 \end{pmatrix} = \begin{pmatrix} L_o \\ \vdots \\ L_r \end{pmatrix}$$

here L_i is a v x (r+1)v matrix. Using the information above, heck that

$$L_i L_j^T = (r-1)\mu J$$

f $i \neq j$ and

$$L_i L_i^T = k(r-\lambda)I + k\lambda J.$$

ow, finally, $(r-1)\mu = (r-1)k^2/v = k\lambda$. Hence

$$LL^T = k(r-\lambda)I + k\lambda J.$$

is then the incidence matrix of the desired symmetric esign.□

Remark. Actually, there are many different ways o construct L. If we set $M_0 = 0$, then we see that the atrix L above is constructed by taking the group operation able for the additive group of integers modulo (r+1) and eplacing i by M_i. In fact, the group table of any group of rder (r+1) serves equally well, if we put the group elements nto one-to-one correspondence with the matrices M_i. (We hall make use of this additional generality in Chapter 4.)

Going a step further, we can use any Latin square f size (r+1) for the construction. (A Latin square of size is an m x m array of m symbols such that every symbol occurs exactly once in each row and column.)

Example 14. We can insert our two classes of affine designs into Theorem 1.13 to obtain new symmetric designs. There exist symmetric (v,k,λ) designs with

(1) $v=q^{d+1}(q^d+\ldots+q+2)$, $k=q^d(q^d+\ldots+q+1)$, $\lambda=q^d(q^{d-1}+\ldots+q+1)$ for every prime power q, and

(2) $v=16n^2$, $k=2n(4n-1)$, $\lambda=2n(2n-1)$ whenever a Hadamard design with parameters $(4n-1, 2n-1, n-1)$ exists.

§1.5 DEMBOWSKI-WAGNER THEOREM

In this section we give most, but not all, of the proof of the Dembowski-Wagner Theorem. The final step requires a celebrated classification of projective geometries, found by Veblen and Young in 1910. To prove Veblen and Young's result in full generality would be too much of a digression (although I hope the reader will want to look it up). However, we can give a quite short proof of Veblen and Young's result for the special case of projective geometries over $\mathbb{F}_2$, which provides a bit of the flavour of the general case.

We begin with a lemma.

Lemma 1.14. *In a symmetric* (v,k,λ) *design*

(1) *given any two points, there is a unique line containing them;*

(2) *a block meets a line in* 0,1 *or all of the line's points;*

(3) *a line has at most* $(v-\lambda)/(k-\lambda)$ *points, with equality if and only if it meets every block;*

(4) *if all lines have equally many, say* h, *points then* $h\leq k$ *(with equality only for* $\lambda=1$*) and the points and lines form a block design with parameters* $(v', b', k', r', \lambda')$, *where*

$$v' = v \qquad b' = v(v-1)/h(h-1)$$
$$k' = h \qquad r' = (v-1)/(h-1)$$
$$\lambda' = 1$$

Proof. (1), (2) and (4) are straightforward. (3) follows from (2).□

Theorem 1.7. (Dembowski-Wagner) Let D be a non-trivial symmetric (v,k,λ) design. The following are equivalent:

(1) Every line has exactly $h=(v-\lambda)/(k-\lambda)$ points.

(2) Every line has at least $h=(v-\lambda)/(k-\lambda)$ points.

(3) Every line meets every block.

(4) The number of blocks through three noncollinear points is constant.

(5) D is either a projective plane or is isomorphic to some PG(m,q) (where $q=h-1$).

Proof. If D is isomorphic to PG(m,q) then, by the remarks in §1.3, properties (1),(2),(3) and (4) hold. Also, if D is a projective plane then the lines are just the blocks and again the properties hold.

Next we show that the first three properties are equivalent to one another. Clearly (1) implies (2). Now if L is a line of D we compute that there are $(v-\lambda)-|L|(k-\lambda)$ blocks which do not intersect L. Hence (2) implies (3) and (3) implies (1).

Since (1),(2) and (3) are equivalent, let us suppose them all and show that (4) follows. Let x,y and z be three noncollinear points and let L be the line through y and z. Suppose that there are ρ blocks containing x,y and z. Let us count the number F of blocks containing x but not L. Certainly $F=k-\rho$. But, if a block contains two points of L it contains all of L, and we have assumed (by property (3)) that every block meets L in at least one point. Thus every block containing x but not L meets L in exactly one point. Through any point of L there are $\lambda-\rho$ such blocks. Thus $|L|(\lambda-\rho)=F=(k-\rho)$. Since the number of points on a line is constant (by assumption) ρ is independent of x,y and z.

Conversely, let us assume (4) and show that (1) follows. We suppose that any three noncollinear points are contained in ρ blocks. Let x and y be two points and let L be the line through them. By counting in two ways the number of pairs (z,B), where z is a point not on L, and B is a block containing x,y and z, we find that

$$(v-h)\rho = \lambda(k-h),$$

where h is the number of points on L. Since $\lambda \neq \rho$, we have

$$h = \frac{\lambda k - v\rho}{\lambda - \rho} .$$

Thus the number of points on a line is a constant h. Choose a point x. Define a new incidence structure $D^*(x)$ as follows: the points are the lines of D through x, the blocks are the blocks of D through x and incidence corresponds to inclusion. Check that $D^*(x)$ is a 2-design with parameters

$$v^* = (v-1)/(h-1) \qquad r^* = \lambda$$
$$b^* = k \qquad \lambda^* = \rho$$
$$k^* = (k-1)/(h-1)$$

Moreover, the number of points $D^*(x)$ contained in the intersection of two blocks of $D^*(x)$ is $(\lambda-1)/(h-1)$. (Why?) Thus $D^*(x)^{dual}$ is a 2-design and, by Corollary 1.11, we see that $D^*(x)$ must be a symmetric design. Hence $v^*=b^*$ and

$$h = 1 + \frac{k-1}{\lambda} = \frac{v-\lambda}{k-\lambda} .$$

We therefore have shown that (4) implies (1). So, the first four properties are equivalent.

Last of all, we show that together (1),(2),(3),(4) imply (5). If $\lambda=1$, then D is by definition a projective plane and (5) is satisfied. So, assume that $\lambda>1$. Let h be the number of points on any line and let ρ be the number of blocks containing any three noncollinear points. (Since $\lambda>1$ then $\rho>0$.) If x,y and z are three noncollinear points, define the <u>plane</u> E through x,y and z to be the intersection of all blocks containing the three points. If u and v are two points of E then the line through u and v lies entirely in E. (Why?) Define an incidence structure on the points of E. We claim first of all that this gives E the structure of a 2-($|E|$,h,1) design. For, there are h points on any line and two points of E lie in a unique line. What about the intersection of two lines L and M? Certainly L and M meet in at most one point. Also, we can find a block B which contains L but not E. Since every block meets M, we have $\emptyset \neq (B\cap M)\cap E = L\cap M$. So, L and M meet in exactly one point. By Corollary 1.11, the incidence structure on E is a symmetric design; in particular, it is a projective plane.

We have enough information now to invoke the classic theorem of Veblen and Young.

Theorem 1.16. (Veblen and Young) Suppose that a collection of subsets (called 'lines') of a finite set of points satisfies the following conditions:

(a) any line contains at least three points, and no line contains every point;

(b) any two points lie on a unique line;

(c) any three noncollinear points lie in a subset which, together with the lines it contains, forms a projective plane.

Then the lines are in fact the lines of a projective geometry or a projective plane.

Moreover, define a subspace of the finite set to be a collection S of the points with the property that if p and q lie in S then the line through p and q lies in S. Then the subspaces correspond exactly to the subspaces of the projective geometry or projective plane. In particular, the maximal proper subspaces correspond to the hyperplanes.

The theorem now completes the proof of the Dembowski-Wagner theorem. For, a symmetric design with properties (1),(2),(3),(4) satisfies Theorem 1.16. (Condition (b) is satisfied since $(v-\lambda)/(k-\lambda) = 1 + (k-1)/\lambda \geq 3$, provided that D is nontrivial.) Since the blocks are clearly maximal proper subspaces, they correspond to the hyperplanes of the appropriate projective geometry over some F_q. The number of points on a line is then $h=q+1$.□

We can give a simple proof of Veblen and Young's result in a special case. Suppose that some line ℓ of the incidence structure in question has three points. Every line must then have three points. (For, consider any other line ℓ'. If ℓ and ℓ' intersect, then they lie in a common projective plane and ℓ' must have 3 points. If ℓ and ℓ' do not intersect, find a line which meets both.)

To extract the structure of a projective geometry over F_2, proceed as follows. Define an addition on $P \cup \{\infty\}$ where P is the set of points and ∞ is some new object. If $p, q \in P$ and $p \neq q$, let $p+q$ be the third point on the line through p and q. If $p \in P$, let $p+p=\infty$. Also let $\infty+\infty=\infty$ and $\infty+p=p+\infty=p$ for all $p \in P$. Addition is commutative. To check associativity it suffices to check that $(p+q)+r=p+(q+r)$, for all $p,q,r \in P$. For this calculation it is enough to restrict attention to the seven-point projective plane containing p,q,r. Since there is a unique such symmetric design (up to isomorphism), represented by Figure 1.1, it is a simple matter to verify that addition is associative in the projective plane containing p,q,r.

So, $P \cup \{\infty\}$, has the structure of an elementary abelian 2-group--or, in other words, an F_2-vector space. The zero vector is ∞. The sets $B \cup \{\infty\}$, where B is a block, are maximal subgroups--that is, subspaces of codimension 1. By discarding ∞, the points and blocks of D have the structure of PG(m,2) for some m.

PROBLEMS - CHAPTER 1

. Show that every automorphism of a symmetric design also ives an automorphism of the complementary design.

. Show that between a symmetric design and its complement xactly one has the property that $v \geq \frac{1}{2}k$ and $n > \lambda$.

A <u>triangle</u> is a set of three points which do not ie on a common block; an <u>ordered triangle</u> is an ordered riple (p_1, p_2, p_3) such that $\{p_1, p_2, p_3\}$ is a triangle.

. Show that any symmetric (7,3,1) design D is isomorphic to G(2,2). In fact, show the following: if (p_1, p_2, p_3) is an rdered triangle in D, and (q_1, q_2, q_3) is an ordered triangle n PG(2,2), we may assign $p_i \longmapsto q_i$ (i=1,2,3) and, once the ssignment is made, we can uniquely extend it to an iso- orphism. As a consequence, show that the full automorphism roup of PG(2,2) is sharply transitive on ordered triangles. Recall that a permutation group G on a set X is <u>sharply</u> <u>ransitive</u> if for any $x, y \in X$ there is exactly one element of sending x to y.) By counting ordered triangles, show that he full automorphism group of PG(2,2) has order 168.

Of course, we already know the order of the full utomorphism group since it is PGL(3,2). However, Problem 3 s an example of the combinatorial principle that "a strong nough uniqueness theorem gives the order of the automorphism roup." Another example is the following:

. Show that any symmetric (11,5,2) design D is isomorphic o H(11). Show that the full automorphic group of H(11) has rder 660. (Hint: count (p_1, p_2, p_3) where p_1, p_2, and p_3 are istinct points of a block B.) Check that the full auto- orphism group acts doubly transitively on points and on locks. How does the subgroup stabilizing a block B act on he points of B?

5. Find some automorphisms of the symmetric design $H'(2q+1)$. Suppose that q and $2q+1$ are both prime powers. By using Theorem 1.5 and some of the automorphisms you have found, show: $H(2q+1)$ and $H'(2q+1)$ are isomorphic if and only if $q=5$.

6. The purpose of this problem is to prove the following statement: if y^2+c is a quadratic with coefficients in F_q (with q odd) and χ is the Legendre symbol of F_q, then

$$\sum_{y \varepsilon F_q} \chi(y^2+c) = \begin{cases} -1 & \text{if } c \neq 0 \\ q-1 & \text{if } c=0. \end{cases}$$

The statement is easy if $c=0$, so suppose that $c \neq 0$.

(i) Let $S(c)$ denote the sum. Show that $S(c)$ depends at worst only on whether c is a square. Show, by reversing the order of summation, that

$$\sum_{c \varepsilon F_q} S(c) = 0.$$

and thus, if d is a nonsquare, that $S(1) + S(d) = -2$.

(ii) Show that $S(1) = -1$. (Hint: in the sum $\Sigma\chi(y^2-1)$ the term corresponding to $y=1$ is zero; exclude it. Let $z = (y+1)/(y-1)$. The summand becomes $\chi(z)\chi(y-1)^2$. As y runs over all elements except 1, so does z.) Hence show that $S(d) = -1$.

7. If ay^2+by+c is a quadratic in F_q (with q odd) and $a \neq 0$, show that

$$\sum_{y \varepsilon F_q} \chi(ay^2+by+c) = \begin{cases} -\chi(a) & \text{if } \Delta \neq 0 \\ (q-1)\chi(a) & \text{if } \Delta \neq 0 \end{cases}$$

where $\Delta = b^2-4ac$. As a corollary, note that

$$\sum_{y \varepsilon F_q} \chi(y+b)\chi(y+c) = \begin{cases} -1 & \text{if } c \neq b \\ q-1 & \text{if } c=b. \end{cases}$$

and use this to prove Proposition 1.4.

8. Verify the details in Example 9.

9. Let D be a symmetric (v,k,λ) design and let B be a block. Prove that if $n>\lambda$ then D^B has no repeated blocks.

10. Use Proposition 1.7 to show that if a t-(v,k,λ) design with b blocks is extendable, then k+1 divides b(v+1). Deduce that if a projective plane of order n is extendable then n=2, 4, or 10.

11. (i) Show that the number of t-dimensional subspaces of an m-dimensional vector space over F_q is

$$\frac{(q^m-1)(q^m-q)\ldots(q^m-q^{t-1})}{(q^t-1)(q^t-q)\ldots(q^t-q^{t-1})}.$$

for $t\leq m$. (Hint: the numerator counts the number of ways to pick an ordered basis for the subspace.)

(ii) How many t-dimensional subspaces of the vector space contain a given h-dimensional subspace (with $h\leq j$)?

(iii) Show that the points and projective t-dimensional subspaces of a projective space over F_q actually form a block design. Compute the parameters v,b,k,r and λ for $P_t(m,q)$.

12. (i) Suppose that G is t-homogeneous permutation group on a set X. (Recall that this means that G transitively permutes the collection of subsets of X of size t.) If S is a subset of X, show that the images of S under G define the blocks of a t-design (which may be the complete t-design) on X.

(ii) If a symmetric design is a 3-design, show that it must be a trivial symmetric design.

(iii) Show that an automorphism group of a nontrivial symmetric design cannot act 3-homogeneously on points.

13. Suppose that the full automorphism group of a nontrivial symmetric design D is transitive on non-collinear triples. Show that D is either a projective plane or is isomorphic to some PG(m,q). (In light of the previous problem, this is as close to 3-transitivity as one can reasonably expect.)

14. Recall the symmetric design A(2m,q) defined in Example 12 Using Appendix B, show that

(i) A(2m,q) is in fact a symmetric design with the same parameters as PG(2m-1,q);

(ii) the full automorphism group of A(2m,q) is transitive on points;

(iii) the line through any two points has two or q+1 points;

(iv) A(2m,q) is not isomorphic to PG(2m-1,q).

15. Let D be an affine 2-(v,k,λ) design with parallel classes of size s and nonparallel blocks meeting in μ points.

(i) Show that $s=v/k=b/r$.

(ii) Fix a block B. Count in two ways pairs (x,B') where B' is a block other than B and x is a point on B and B'. Conclude that $\mu=k/s=v/s^2$.

(iii) Show then that $v=\mu s^2$, $k=\mu s$, $\lambda=(\mu s-1)/(s-1)$, $r=(\mu s^2-1)/(s-1)^2$, and $b=s(\mu s^2-1)/(s-1)$. Note that $\lambda-\mu=(\mu-1)/(s-1)$ and thus that $(s-1)$ divides $(\mu-1)$.

16. Recall that the residual design of PG(m,q) is AG(m,q) and that the derived design is q copies of PG(m-1,q). This problem generalizes this situation by concentrating on the affine design as a link between two symmetric designs. Let D be an affine 2-design with parameters as in the previous problem.

(i) Suppose that there exists a symmetric (V,K,Λ) design P such that P^W is isomorphic to D, for some block W. We have $V = b+1$, $K=r$ and $\Lambda=\lambda$. Show that P_W has the structure of a symmetric (v^*,k^*,λ^*) design, except that each block is repeated s times (corresponding to each of the elements in the parallel class), where

$$v^*=r,\ k^*=\lambda \text{ and } \lambda^*=\lambda-\mu.$$

(ii) Conversely, suppose that there exists a symmetric (v^*,k^*,λ^*) design P^* with $v^*=r$, $k^*=\lambda$ and $\lambda^*=\lambda-\mu$. Given any one-to-one correspondence between parallel classes of D and blocks of P^*, construct a symmetric (V,K,Λ) design P with $V=b+1$, $K=r$ and $\Lambda=\lambda$ such that P^W is isomorphic to D and P_W is isomorphic to s copies of P^*, for some block W.

(iii) Suppose we apply the construction in (ii) to AG(m,q), using $P^* = PG(m-1,q)$. Observe that the resulting symmetric design need not be isomorphic to PG(m,q), depending on how the one-to-one correspondence is chosen. One can guarantee that P is not isomorphic to PG(m,q) by choosing the correspondence in such a way that some line of P does not meet the block of P corresponding to P^*.

17. Show that, for any positive integers t,v,k (with $t<k<v-t$) there is some integer λ for which there exists a t-(v,k,λ) design (possibly with repeated blocks) in which not every point set of cardinality k occurs as the set of points incident with a block. (Hint: consider the "incidence matrix" of point sets of cardinality t versus those of cardinality k. Notice that the columns are linearly dependent over the rationals.)

18. Let D be a t-$(v,k,1)$ design, let B be a block of D and let $I= \{p_1,\ldots,p_i\}$ be a set of i points in B (with $i\leq k$). Let λ_{ij} be the number of blocks B' such that $B'\cap\{p_1,\ldots,p_i\} = \{p_1,\ldots,p_j\}$ for $j\leq i$. The point of this problem is to show that λ_{ij} depends only on i,j,t,v, and k, but not on the choice of blocks or points.

(i) Show that λ_{ii} is well defined and equal to one.

(ii) Show that the numbers λ_{ij} must satisfy the Pascal property

$$\lambda_{ij} = \lambda_{i+1,j} + \lambda_{i+1,j+1}$$

and thereby simultaneously prove by induction on i-j that the λ_{ij} are well defined and provide a formula for computing them.

(iii) As an example, the <u>intersection triangle</u> of a 5-(24,8,1) design, which we shall construct in the next chapter, is

759								
506	253							
330	176	77						
210	120	56	21					
130	80	40	16	5				
78	52	28	12	4	1			
46	32	20	8	4	0	1		
30	16	16	4	4	0	0	1	
30	0	16	0	0	0	0	1	

Observe, incidentally, that distinct blocks must meet in 0, 2 or 4 points.

(iv) As a slight variation on this theme, prove the following result. Let B and I be as above and let x be a point not on B. Then the number of blocks of B' such that $x \in B'$ and $B' \cap B = I$ depends only on i,t,v, and k. Show that this number is $\mu_i = \lambda_{ki}(k-i)/(v-k)$.

upplementary Problems: Algebraic Geometry

These supplementary problems together prove the ollowing result, alluded to in §1.3.

Theorem 1.17. The lines of $H(q)$ have size 2 if $q>7$.

Since the lines of $PG(m,2)$ have size 3, this shows hat $H(q)$ and $PG(m,2)$ are not isomorphic unless $(q,m)=(7,2)$. lthough the proof is a little involved I have included it ecause it introduces the reader, in an elementary way, to ome algebraic geometry and shows how it can be exploited n combinatorics.

The lines of $H(q)$ have cardinality at most 3 by emma 1.14. If one line has size 3 then all lines have ize 3, since $H(q)$ has a 2-homogeneous automorphism group. o, for the purpose of obtaining a contradiction, suppose hat all lines of $H(q)$ have size 3.

). (i) Recall the automorphism group $S(q)$, which consists f transformations of the form $x \longmapsto ax + b$, where a is nonzero square and b is any element of F_q. Show that only he identity element fixes more than 2 points.

(ii) Show that the points and lines form a block design ith $v=q$, $b=(q^2-q)/6$, $k=3$, $r=(q-1)/2$ and $\lambda=1$. The group (q) permutes the lines transitively.

(iii) Show that the subgroup of $S(q)$ stabilizing a line (setwise) has order 3 and permutes the points of L in a -cycle.

). Suppose that q is not a power of 3.

(i) Show that 6 divides $q-1$ and hence that F_q contains primitive cube root of unity, ζ.

(ii) Let L be the line containing 1 and ζ. Show that he subgroup stabilizing L is generated by $x \longmapsto \zeta x$ and hat $L = \{1, \zeta, \zeta^2\}$.

(iii) Whenever two points of L are on a block, so is the hird. Hence for $x \in F_q$, if any two of $x+1$, $x+\zeta$ and $x+\zeta^2$

are nonzero squares then so is the third. The $(x+1)(x+\zeta)(x+\zeta^2)$ is a square, except when all three of the factors are nonsquares.

(iv) Show that the equation $y^2=x^3-1$ has at least $\frac{3}{2}(q-1)$ distinct solutions (x,y) with $x, y \in \mathbb{F}_q$. (Remember to count two solutions when x^3-1 is a nonzero square.)

21. Suppose now that q is a power of 3.

(i) Argue similarly to show that the third point on the line through 0 and 1 is -1.

(ii) Show that the equation

$$y^2=x^3-x$$

has at least $\frac{3}{2}(q-1)$ solutions.

It seems rather unlikely that such equations would have so many solutions. Naively, there is no reason to suppose that x^3-1 is either a square or a nonsquare more often than not. If it is a square (and so accounts for two solutions to $y^2=x^3-1$) for about half the values of x, then one would expect approximately q solutions (x,y) to $y^2=x^3-1$. In fact, this naive estimate is quite good.

Algebraic geometry provides estimates for the number of solutions to equations over finite fields. For example, it is known that the number M of solutions to either $y^2=x^3-1$ or $y^2=x^3-x$ in $\mathbb{F}_q$ satisfies the inequality

$$|M-q| \leq 2\sqrt{q}\ . \qquad (*)$$

Thus these equations could only have $\frac{3}{2}(q-1)$ or more solutions over very small fields.

The inequality (*) is a special case of a much more general estimate due to Weil for the number of points on an algebraic curve. Let $f(x,y)$ be an absolutely irreducible polynomial over $\mathbb{F}_q$ (i.e., $f(x,y)$ does not factor over any algebraic extension of $\mathbb{F}_q$). The number N of solutions $(x,y) \in \mathbb{F}_q \times \mathbb{F}_q$ to $f(x,y) = 0$ satisfies

$$|N-q| \leq 2g\sqrt{q}$$

where g is the "genus" of the curve $f(x,y)=0$. (Equations of the form $y^2=f(x)$, where $f(x)$ is a cubic polynomial, give rise to elliptic curves, which have genus 1.) While Weil's estimate is not easy to prove, even for elliptic curves, the two equations $y^2=x^3-1$ and $y^2=x^3-x$ are rather special. In these cases, we can obtain estimates of the number of solutions by entirely elementary methods, due to Jacobsthal.

22. Suppose that q is an odd prime power congruent to 1 (mod 3). Let χ be the Legendre symbol of F_q. For $a\varepsilon F_q$, define the quantity

$$S(a) = \sum_{x\varepsilon F_q} \chi(x^3 + a).$$

(i) Show that $S(0)=0$.

(ii) Show that $|S(a)| = |S(ay^3)|$ for all $y \varepsilon F_q-\{0\}$. Thus, for $a\neq 0$, the expression $|S(a)|$ assumes at most three values, depending only on the coset of the subgroup of cubes which contains a. If α is some non-cube, the values are $S(1)$, $S(\alpha)$ and $S(\alpha^2)$.

(iii) By interchanging the order of summation show that

$$\sum_{a\varepsilon F_q} S(a) = 0$$

(iv) Consider the sum

$$\sum_{a\varepsilon F_q} S(a)^2 = \sum_{a\varepsilon F_q} \sum_{x\varepsilon F_q} \sum_{z\varepsilon F_q} \chi(x^3+a)\chi(y^3+a)$$

$$= \sum_{x\varepsilon F_q} \sum_{y\varepsilon F_q} [\sum_{a\varepsilon F_q} \chi((a+x^3)(a+y^3))].$$

By using Problem 7, evaluate the inner summation and show that

$$\sum_{a\varepsilon F_q} S(a)^2 = 2q(q-1).$$

Hence $S(1)^2 + S(\alpha)^2 + S(\alpha^2)^2 = 6q$. Thus $|S(a)| \leq \sqrt{6q}$ for all $a \in F_q$.

(v) Let M be the number of solutions to $y^2=x^3-1$. Show that $M-q=S(-1)$. Hence $|M-q| \leq \sqrt{6q}$, giving us an estimate of the number of solutions and a bound on the error.

(vi) Show that $M<\frac{3}{2}(q-1)$ except possibly if $q\leq 29$. Thus, when $q\equiv 1 \pmod 3$, the lines of H(q) have size 2 except possibly if q=7,13,19 or 25. Exclude the last three cases by a direct check.

23. Suppose that q is a power of 3. For $a\in F_q$, let

$$T(a) = \sum_{x\in F_q} \chi(x^3 + ax)$$

(i) Show that $T(0) = 0$.

(ii) Show that $|T(a)| = |T(ay^2)|$ for all $y\in F_q-\{0\}$ and thus that $|T(a)|$ assumes at most two values, T(1) and T(d), where d is a nonsquare. (Hint: $(u^3 + ay^2u) = y^3(x^3 + ax)$ where $u = yx$.)

(iii) Show that

$$\sum_{a\in F_q} T(a)^2 = q(q-1)(1+\chi(-1)).$$

and thus that $|T(a)| \leq \sqrt{2(1+\chi(-1))q}$ for all $a \in F_q$. In particular when q is an odd power of 3, then $T(a)=0$ for all $a\in F_q$.

(iv) Let M be the number of solutions of $y^2=x^3-x$. Show that $M-q = T(-1)$.

(v) The design H(q) is defined only for q an odd power of 3. In these cases, $M=q$. Hence $M<\frac{3}{2}(q-1)$ unless $q=3$. This completes the proof that the lines of H(q) have size 2 whenever $q>7$.

NOTES TO CHAPTER 1

§1.1 Combinatorial questions related to designs ere studied by Steiner, Kirkman, E.H. Moore and others in ne nineteenth century. Fisher and Yates introduced the otion of a "balanced design"--which corresponds to what is ow known as a 2-design--in connection with the statistical neory of experimental design. Symmetric (v,k,λ) designs re sometimes referred to as "projective designs," or ccasionally, as "λ-planes."

The reader is warned that a different convention for efining incidence matrices is sometimes used, for example in ne books by Dembowski [33] and Hall [49], with the result that he incidence matrices there are the transpose of ours.

Another symmetric (16,6,2) design may be constructed s follows. Take a 4 x 4 Latin square, say

a	b	c	d
d	a	b	c
c	d	a	b
b	c	d	a

nd, for each entry, take for a block corresponding to that ntry the six entries not in its row or column and not with ne same label. So, e.g., the block corresponding to (1,1) is

$$\{(2,3),(2,4),(3,2),(3,4),(4,2),(4,3)\}$$

he resulting design is not isomorphic to that of Example 4; n particular, the incidence matrices of the two designs have ifferent ranks modulo 2. (Cf. §2.2.)

§1.2 Hadamard [45] showed: if $H=(h_{ij})$ is an m x m atrix with $-1\leq h_{ij}\leq 1$ for $1\leq i,j\leq m$ then $|\det H| \leq m^{\frac{1}{2}m}$, with equality if and only if H is a (-1,1)-matrix satisfying $HH^T=mI$. he constructions using Jacobsthal matrices are due to Paley 109]. Concerning Hadamard designs, see Wallis, Street, and

Wallis [138], Geramita and Seberry [40] and Seberry [121].

§1.3 Example 13 seems to have been discovered independently by several authors. See, e.g., Cameron and van Lint [25, p. 48].

§1.4 The standard reference work on t-designs (and many other topics) is Dembowski [33]. Much interest in the connection between groups and t-designs was stimulated by Hughes [63].

Some 5-designs (without repeated blocks) have been constructed by Alltop [2], Assmus and Mattson [6], Denniston [35] and Pless [112]. Theorem 1.12 is due to Wallis [137], but we give the proof of Lenz and Jungnickel [83]. Part (1) of Example 14 was found by Ahrens and Szekeres [1] in the case d=1 and by McFarland [97] in general.

§1.5 Lines can be defined analogously for arbitrary 2-designs, and, with two changes, the Dembowski-Wagner Theorem remains true for arbitrary 2-designs. First, replace $(v-\lambda)/(k-\lambda)$ by $(b-\lambda)/(r-\lambda)$ in (1) and (2). Second, add to (4) the hypothesis that D is a symmetric design. (Clearly some condition must be added to (4) since affine designs, for example, always have a constant number of blocks containing three non-collinear points.)

Tsuzuku's recent book [131] contains an accessible exposition of Veblen and Young's theorem. The Veblen-Young result [136] is the grandfather of many later theorems which characterize incidence structures by a few axioms. See Buekenhout [21] (affine geometries), Buekenhout and Shult [22](polar spaces), and Shult and Yanushka [125] and Cameron [27] (dual polar spaces).

Problems. Problem 10 is due to Hughes [63]. Cameron [24] gives the best answer yet to the question: when is a symmetric (v,k,λ) design extendable? Problem 12 appears in Dembowski [33, p. 5]. Concerning Problem 16, we can always reconstruct PG(m,q) from its residual design, the affine geometry. We can ask more generally: suppose that a 2-(v,k,λ) design has the right parameters to be the residual of a symmetric design with respect to a block; is it? The answer is Yes if k is sufficiently large with respect to λ. See Hall and Connor [52] and Bose, Shrikhande and Singhi [18]. Concerning Problem 18, for more on intersection numbers, see Cameron [26].

Supplementary Problems. Problems 19, 20 and 21 are due to Lander. I am indebted to B.J. Birch for Problems 22 and 23 which he credits to Jacobsthal. The combinatorialist interested in learning more about Weil's estimate is recommended to the monograph of Schmidt [119], which contains an elementary proof using the method of Stepanov. Hirschfeld [59] provides examples of applications to combinatorics.

Problem 23 proves that the exact number of points on the curve $y^2=x^3-ax$ over $\mathbb{F}_q$ is q, whenever q is an odd power of 3. J.E. Cremona points out that when q is an even power of 3 the number of points can also be determined exactly; it is q or $q-2\sqrt{q}$, according as 2 is a square or non-square.

2. AN ALGEBRAIC APPROACH

§2.1 EXISTENCE CRITERIA

The most fundamental question one can ask about our topic is: which triples (v,k,λ) are the parameters of a symmetric design? The question is far from resolved. Certainly, (v,k,λ) must satisfy $(v-1)\lambda = k(k-1)$ and the conclusion of Schutzenberger's Theorem. A deeper condition is given by the Bruck-Ryser-Chowla Theorem:

Theorem 2.1. *Suppose that a symmetric* (v,k,λ) *design exists. If* v *is odd then the equation*

$$nX^2 + (-1)^{\frac{v-1}{2}} \lambda Y^2 = Z^2$$

must have a solution in integers X,Y,Z, *not all zero*.

Remark: We may drop the condition that v is odd, if we choose. For, if v is even, n must be a square and $(1,0,\sqrt{n})$ solves the equation in integers.

Proof. Let A be the incidence matrix of a symmetric (v,k,λ) design. Define two square matrices, each of order $(v+1)$:

$$B = \left(\begin{array}{ccc|c} & A & & \begin{matrix}1\\ \vdots \\ 1\end{matrix} \\ \hline \lambda & & \lambda & k \end{array}\right) \quad \text{and} \quad \psi = \begin{pmatrix} 1 & & & 0 \\ & \ddots & & \\ & & 1 & \\ 0 & & & -\lambda \end{pmatrix} .$$

Using the block-intersection properties, check that

$$B\psi B^T = n\psi \ .$$

Interpret this matrix equation as a statement about the equivalence of rational quadratic forms: it says that the quadratic forms $Q_1 \equiv y_1^2 + \ldots + y_v^2 - \lambda y_{v+1}^2$ and $Q_2 \equiv nx_1^2 + \ldots + nx_v^2 - n\lambda x_{v+1}^2$ are equivalent.

Lemma 2.2. For any positive integer t, the quadratic forms $y_1^2 + y_2^2 + y_3^2 + y_4^2$ and $tx_1^2 + tx_2^2 + tx_3^2 + tx_4^2$ are equivalent.

We postpone the proof for a moment. Using the lemma and Witt's Cancellation Theorem (Appendix B), we can cancel terms, four at a time, from the equivalent quadratic forms Q_1 and Q_2 above.

Case 1. $v \equiv 1 \pmod 4$. After cancellation, we find that $y_v^2 - \lambda y_{v+1}^2$ and $nx_v^2 - n\lambda x_{v+1}^2$ are equivalent quadratic forms. Hence they represent the same numbers. The latter represents n (at the point (1,0)). Hence there exists a rational point (z,y) such that

$$z^2 - \lambda y^2 = n$$

Multiplying through by a common denominator, we find an integral solution to the equation in the theorem.

Case 2. $v \equiv 3 \pmod 4$. Add the quadratic form $u^2 + nv^2$ to each of the quadratic forms. The resulting quadratic forms in (v+3) variables are equivalent. Proceed as before. (Check the details.)

Finally, we prove the lemma.

Proof of lemma. We need two facts from number theory, both due to Lagrange. The first is the famous 'four square' identity:

$$(a_1^2+a_2^2+a_3^2+a_4^2)(x_1^2+x_2^2+x_3^2+x_4^2)=(y_1^2+y_2^2+y_3^2+y_4^2),$$

where
$$y_1 = a_1x_1 - a_2x_2 - a_3x_3 - a_4x_4,$$
$$y_2 = a_1x_2 + a_2x_1 + a_3x_4 - a_4x_3,$$
$$y_3 = a_1x_3 + a_3x_1 + a_4x_2 - a_2x_4,$$
$$y_4 = a_1x_4 + a_4x_1 + a_2x_3 - a_3x_2.$$

Interpret the x_i and y_i as formal variables and the a_i as integers. The identity provides us with an explicit transformation changing the quadratic form $t(x_1^2+x_2^2+x_3^2+x_4^2)$ into $(y_1^2+y_2^2+y_3^2+y_4^2)$, where t is the integer $t=a_1^2+a_2^2+a_3^2+a_4^2$. By another theorem of Lagrange, every positive integer can be written as the sum of four squares. Together the results of Lagrange imply the lemma.□

Remark. The matrix B, defined in the proof, is called the extended incidence matrix of the symmetric design and it will be our mainstay throughout this chapter. In view of the equation $B\psi B^T=n\psi$, we have $|\det B| = n^{\frac{1}{2}(v+1)}$.

As an application, consider projective planes. Here $\lambda=1$ and $v=n^2+n+1$ is odd. If $n\equiv 0$ or 3 (mod 4), the Bruck-Ryser-Chowla equation always has the solution (0,1,1) and thus the theorem excludes no values of n. However, if $n\equiv 1$ or 2 (mod 4), the equation becomes $nx^2 = y^2 + z^2$, which has a nontrivial integral solution if and only if n is the sum of two squares of integers. (See Problem 1.) Projective planes of order 6,14,21,22,30 or 33 therefore cannot exist.

Despite much research no one has uncovered any further necessary conditions for the existence of a symmetric (v,k,λ) design apart from the equation $(v-1)\lambda=k(k-1)$, Schutzenberger's Theorem and the Bruck-Ryser-Chowla Theorem. for no (v,k,λ) satisfying these requirements has it been shown that a symmetric (v,k,λ) design does not exist.

It is possible that these conditions are sufficient. As a matter of fact, this is true for the seventeen admissible (v,k,λ) with $v\leq 48$; the first open case as of early 1982 is (49,16,5). On the other hand, a wide gulf separates the list of admissible (v,k,λ) from the list of those for which a symmetric (v,k,λ) design is known to exist. Consider:

- Projective planes of order n exist for all prime powers n (aside from PG(2,n) a host of other constructions are known) but for no other n is a construction known. The first open values are n=10,12,15,18,20,24,26 and 28.

- For other symmetric designs, the situation is even more extreme. For each $\lambda>1$, only finitely many symmetric (v,k,λ) designs are known. In fact, in all but four known nontrivial cases $v\leq\lambda^2(\lambda+2)$. The four exceptions are (37,9,2), (56,11,2), (79,13,2) and (71,15,3). (For each prime power λ, Example 14 supplies a symmetric design attaining the bound $v=\lambda^2(\lambda+2)$.)

Two incompatible conjectures suggest themselves:

Conjecture 2.3. If v,k,λ are positive integers atisfying $(v-1)\lambda=k(k-1)$ and the conditions of Schutzenberer's heorem and the Bruck-Ryser-Chowla Theorem then there exists symmetric (v,k,λ) design.

Conjecture 2.4. For every $\lambda>1$, there exist (up to somorphism) only finitely many symmetric (v,k,λ) designs.

Before leaving the topic of the Bruck-Ryser-Chowla onditions, let us explore how to tell when an equation such s that in the B-R-C Theorem has a nontrivial integral olution. Consider the equation

$$Ax^2 + By^2 + Cz^2 = 0, \qquad (*)$$

nd assume initially that A,B, and C are square-free integers, airwise relatively prime. Suppose that (x,y,z) is a non-rivial integral solution. If p is an odd prime dividing A, e may assume (after possibly dividing our solution through y a power of p) that $p \nmid y$ and $p \nmid z$. Then, $By^2 \equiv -Cz^2 \pmod p$ nd $-BC$ must be a square (mod p). Necessary conditions for he existence of a nontrivial integral solution therefore re that, for all odd primes p,

(1) If $p|A$, then $-BC$ is a square (mod p),

(2) If $p|B$, then $-AC$ is a square (mod p),

(3) If $p|C$, then $-AB$ is a square (mod p),

nd, of course,

(4) A,B, and C do not all have the same sign.

It is a classical theorem, due to Legendre that hese simple necessary conditions are sufficient.

If A,B, and C do not satisfy our assumptions above e may slightly modify the equation (*). Henceforth, let m^* denote the square-free part of the integer m. Then (*) as a nontrivial integral solution if and only if

$$A^*x^2 + B^*y^2 + C^*z^2 = 0$$

as a nontrivial integral solution. Also, if p divides all hree coefficients; we may divide it out and if p divides nly A and B, then (*) has a nontrivial integral solution f and only if

$$(\frac{A}{p})x^2 + (\frac{B}{p})y^2 + (pC)z^2 = 0$$

does. Hence (*) can always be transformed into an equation to which Legendre's result applies. By applying Legendre's theorem to the Bruck-Ryser-Chowla equation, we obtain an alternate version of Theorem 2.1, which is much easier to apply in practice.

Theorem 2.5. Suppose that there exists a symmetric (v,k,λ) design. Then for every odd prime p,

(1) If $p \nmid n^*$ and $p \mid \lambda^*$, then n is a square (mod p).

(2) If $p \mid n^*$ and $p \nmid \lambda^*$, then $(-1)^{\frac{1}{2}(v-1)}\lambda^*$ is a square (mod p).

(3) If $p \mid n^*$ and $p \mid \lambda^*$, then $(-1)^{\frac{1}{2}(v+1)}(\lambda^*/p)(n^*/p)$ is a square (mod p).

The aim of the rest of this chapter will be to provide a different perspective on the Bruck-Ryser-Chowla Thoerem. At the moment, the three number-theoretic criteria of Theorem 2.5 are without much independent meaning. By the end of the chapter, however, we shall see that conditions (2) and (3) are formally equivalent to asserting the existence of an object, called a self-dual code, associated with certain symmetric designs. These self-dual codes will provide us with a powerful new tool which we shall exploit in later chapters.

§2.2 THE CODE OF A SYMMETRIC DESIGN

The rows of the incidence matrix of a symmetric (v,k,λ) design can be viewed as v-tuples of ones and zeroes--elements in R^v, the free module of rank v over any ring R we choose. The submodule spanned by these rows turns out to have some rather remarkable properties, many of which have only been recognized quite recently. In this section we concentrate on the case in which R is a finite field $\mathbb{F}_p$ of prime order, p. For this purpose, we adopt the terminology of algebraic coding theory.

A (linear) code C of length m, over a field F, is a subspace of the vector space $V=F^m$. If C has dimension r, we say that C is an [m,r] code. The elements of C are called codewords. The weight of a codeword is the number of nonzero coordinates in it. The minimum weight of a code is the smallest nonzero weight of any codeword. Codes arise in the mathematical theory of communication and are used to guard against errors caused by "noise" in the transmission channel. The codewords in a code C are taken to be the allowable messages in a scheme. A received message, which may have been garbled, is interpreted as the "nearest" codeword. For codes with a minimum weight d, an error in decoding occurs only if at least [d/2] coordinates have been changed in transmission.

We shall often make use of bilinear forms on F^m. If ψ is a symmetric bilinear form (or a scalar product) on F^m, the dual of a code C with respect to ψ is the code

$$C^{\psi} = \{x \in F^m \mid \psi(x,y) = 0 \text{ for all } y \in C\}.$$

A code C is self-orthogonal with respect to ψ if $C \subseteq C^{\psi}$ and self-dual if $C=C^{\psi}$. If ψ is nonsingular, a self-dual code has dimension $\frac{1}{2}m$.

The F_p-code of a symmetric (v,k,λ) design D is the subspace C spanned by the rows of the incidence matrix A of D. For example, if D is the (unique) symmetric (7,3,1) design, with incidence matrix:

$$A = \begin{pmatrix} 1 & 1 & 0 & 1 & 0 & 0 & 0 \\ 0 & 1 & 1 & 0 & 1 & 0 & 0 \\ 0 & 0 & 1 & 1 & 0 & 1 & 0 \\ 0 & 0 & 0 & 1 & 1 & 0 & 1 \\ 1 & 0 & 0 & 0 & 1 & 1 & 0 \\ 0 & 1 & 0 & 0 & 0 & 1 & 1 \\ 1 & 0 & 1 & 0 & 0 & 0 & 1 \end{pmatrix}$$

then the codewords of the $\mathbb{F}_2$-code of D are (0,0,0,0,0,0,0), (1,1,1,1,1,1,1), (1,1,0,1,0,0,0), (0,1,1,0,1,0,0), (0,0,1,1,0,1,0), (0,0,0,1,1,0,1), (1,0,0,0,1,1,0), (0,1,0,0,0,1,1), (1,0,1,0,0,0,1), (0,0,1,0,1,1,1), (1,0,0,1,0,1,1), (1,1,0,0,1,0,1), (1,1,1,0,0,1,0), (0,1,1,1,0,0,1), (1,0,1,1,1,0,0), and (0,1,0,1,1,1,0). The code has length 7, dimension 4 and minimum weight 3. The codewords of weight 3 are precisely the characteristic functions of the blocks; those of weight 4 are the characteristic functions of the complements.

The dimension of the $\mathbb{F}_p$-code of a symmetric design D is an invariant associated with the isomorphism class of D. The matrices in Figure 2.1 are incidence matrices of symmetric (16,6,2) designs. With patience, we might calculate that the dimension of the $\mathbb{F}_2$-code associated with B_i is i (for i=6,7,8) and thus that no two are isomorphic.

$$B_6 = \begin{pmatrix}
0&1&1&1&1&0&0&0&1&0&0&0&1&0&0&0\\
1&0&1&1&0&1&0&0&0&1&0&0&0&1&0&0\\
1&1&0&1&0&0&1&0&0&0&1&0&0&0&1&0\\
1&1&1&0&0&0&0&1&0&0&0&1&0&0&0&1\\
1&0&0&0&0&1&1&1&1&0&0&0&1&0&0&0\\
0&1&0&0&1&0&1&1&0&1&0&0&0&1&0&0\\
0&0&1&0&1&1&0&1&0&0&1&0&0&0&1&0\\
0&0&0&1&1&1&1&0&0&0&0&1&0&0&0&1\\
1&0&0&0&1&0&0&0&0&1&1&1&1&0&0&0\\
0&1&0&0&0&1&0&0&1&0&1&1&0&1&0&0\\
0&0&1&0&0&0&1&0&1&1&0&1&0&0&1&0\\
0&0&0&1&0&0&0&1&1&1&1&0&0&0&0&1\\
1&0&0&0&1&0&0&0&1&0&0&0&0&1&1&1\\
0&1&0&0&0&1&0&0&0&1&0&0&1&0&1&1\\
0&0&1&0&0&0&1&0&0&0&1&0&1&1&0&1\\
0&0&0&1&0&0&0&1&0&0&0&1&1&1&1&0
\end{pmatrix}$$

Figure 2.1. Incidence matrices of three symmetric (16,6,2) designs

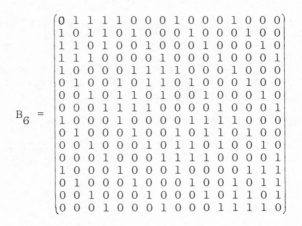

$$
B_7 = \begin{pmatrix}
1 & 1 & 1 & 0 & 0 & 1 & 0 & 0 & 1 & 0 & 0 & 0 & 0 & 0 & 1 & 0 \\
0 & 1 & 1 & 1 & 0 & 0 & 1 & 0 & 0 & 1 & 0 & 0 & 0 & 0 & 0 & 1 \\
0 & 0 & 1 & 1 & 1 & 0 & 0 & 1 & 1 & 0 & 1 & 0 & 0 & 0 & 0 & 0 \\
1 & 0 & 0 & 1 & 1 & 1 & 0 & 0 & 0 & 1 & 0 & 1 & 0 & 0 & 0 & 0 \\
0 & 1 & 0 & 0 & 1 & 1 & 1 & 0 & 0 & 0 & 1 & 0 & 1 & 0 & 0 & 0 \\
0 & 0 & 1 & 0 & 0 & 1 & 1 & 1 & 0 & 0 & 0 & 1 & 0 & 1 & 0 & 0 \\
1 & 0 & 0 & 1 & 0 & 0 & 1 & 1 & 0 & 0 & 0 & 0 & 1 & 0 & 1 & 0 \\
1 & 1 & 0 & 0 & 1 & 0 & 0 & 1 & 0 & 0 & 0 & 0 & 0 & 1 & 0 & 1 \\
1 & 0 & 0 & 0 & 0 & 0 & 1 & 0 & 1 & 1 & 1 & 0 & 0 & 1 & 0 & 0 \\
0 & 1 & 0 & 0 & 0 & 0 & 0 & 1 & 0 & 1 & 1 & 1 & 0 & 0 & 1 & 0 \\
1 & 0 & 1 & 0 & 0 & 0 & 0 & 0 & 0 & 0 & 1 & 1 & 1 & 0 & 0 & 1 \\
0 & 1 & 0 & 1 & 0 & 0 & 0 & 0 & 1 & 0 & 0 & 1 & 1 & 1 & 0 & 0 \\
0 & 0 & 1 & 0 & 1 & 0 & 0 & 0 & 0 & 1 & 0 & 0 & 1 & 1 & 1 & 0 \\
0 & 0 & 0 & 1 & 0 & 1 & 0 & 0 & 0 & 0 & 1 & 0 & 0 & 1 & 1 & 1 \\
0 & 0 & 0 & 0 & 1 & 0 & 1 & 0 & 1 & 0 & 0 & 1 & 0 & 0 & 1 & 1 \\
0 & 0 & 0 & 0 & 0 & 1 & 0 & 1 & 1 & 1 & 0 & 0 & 1 & 0 & 0 & 1
\end{pmatrix}
$$

$$
B_8 = \begin{pmatrix}
1 & 1 & 1 & 1 & 0 & 0 & 0 & 0 & 1 & 0 & 0 & 0 & 1 & 0 & 0 & 0 \\
0 & 1 & 0 & 1 & 1 & 0 & 1 & 0 & 0 & 1 & 0 & 0 & 0 & 1 & 0 & 0 \\
0 & 1 & 1 & 0 & 1 & 0 & 0 & 1 & 0 & 0 & 1 & 0 & 0 & 0 & 1 & 0 \\
0 & 0 & 1 & 1 & 1 & 1 & 0 & 0 & 0 & 0 & 0 & 1 & 0 & 0 & 0 & 1 \\
0 & 0 & 0 & 0 & 1 & 1 & 1 & 1 & 1 & 0 & 0 & 0 & 1 & 0 & 0 & 0 \\
1 & 0 & 1 & 0 & 0 & 1 & 0 & 1 & 0 & 1 & 0 & 0 & 0 & 1 & 0 & 0 \\
1 & 0 & 0 & 1 & 0 & 1 & 1 & 0 & 0 & 0 & 1 & 0 & 0 & 0 & 1 & 0 \\
1 & 1 & 0 & 0 & 0 & 0 & 1 & 1 & 0 & 0 & 0 & 1 & 0 & 0 & 0 & 1 \\
1 & 0 & 0 & 0 & 1 & 0 & 0 & 0 & 1 & 1 & 1 & 1 & 0 & 0 & 0 & 0 \\
0 & 1 & 0 & 0 & 0 & 1 & 0 & 0 & 0 & 1 & 0 & 1 & 1 & 0 & 1 & 0 \\
0 & 0 & 1 & 0 & 0 & 0 & 1 & 0 & 0 & 1 & 1 & 0 & 1 & 0 & 0 & 1 \\
0 & 0 & 0 & 1 & 0 & 0 & 0 & 1 & 0 & 0 & 1 & 1 & 1 & 1 & 0 & 0 \\
1 & 0 & 0 & 0 & 1 & 0 & 0 & 0 & 0 & 0 & 0 & 0 & 1 & 1 & 1 & 1 \\
0 & 1 & 0 & 0 & 0 & 1 & 0 & 0 & 1 & 0 & 1 & 0 & 0 & 1 & 0 & 1 \\
0 & 0 & 1 & 0 & 0 & 0 & 1 & 0 & 1 & 0 & 0 & 1 & 0 & 1 & 1 & 0 \\
0 & 0 & 0 & 1 & 0 & 0 & 0 & 1 & 1 & 1 & 0 & 0 & 0 & 0 & 1 & 1
\end{pmatrix}
$$

Figure 2.1. (cont'd).

Considering blocks of a symmetric design as codewords, the ordinary dot product of two blocks is precisely the number of points the blocks share (reduced mod p). The regularity of block intersection ensures that the code enjoys rather nice properties with respect to the dot product and other bilinear forms. Specifically, let D be a symmetric (v,k,λ) design and let p be a prime which divides n:

(1) If p divides k (and therefore also λ) the dot product of two blocks is always 0 (mod p). Since the F_p-code C of D is generated by these blocks, C is self-orthogonal with respect to the ordinary dot product.

(2) If p does not divide k, we use a slightly different code. The <u>extended F_p-code</u> C^{ext} of D is the F_p-span of the rows of the extended incidence matrix.

$$B = \left(\begin{array}{c|c} & 1 \\ A & \vdots \\ & 1 \\ \hline \lambda\ldots\lambda & k \end{array}\right).$$

Instead of the ordinary dot product we define the bilinear form ψ by

$$\psi(\bar{x},\bar{y}) = x_1y_1+\ldots+x_vy_v - \lambda x_{v+1}y_{v+1},$$

for $\bar{x} = (x_1,\ldots,x_{v+1})$ and $\bar{y} = (y_1,\ldots,y_{v+1})$. Check that if $\bar{x}$ and $\bar{y}$ are rows of B, then $\psi(\bar{x},\bar{y})=0$ or n. Thus $\psi(\bar{x},\bar{y})\equiv 0 \pmod p$. The extended F_p-code C^{ext} is then self-orthogonal with respect to ψ. (N.B. The scalar product ψ is nondegenerate since if $p\nmid k$ then also $p\nmid\lambda$.)

For every prime divisor p of n, we may find an F_p-code (of length v or v+1) which is self-orthogonal with respect to an appropriate nondegenerate scalar product. The next result shows that if $p\nmid n$, the F_p-code is comparatively dull.

<u>Proposition 2.6</u>. <u>Suppose that</u> C <u>is the F_p-code of a symmetric</u> (v,k,λ) <u>design</u>.

(1) <u>If</u> $p|n$, <u>then</u> $2 \le \dim C \le \frac{1}{2}(v+1)$.

(2) <u>If</u> $p\nmid n$, <u>and</u> $p|k$, <u>then</u> $\dim C=v-1$.

(3) <u>If</u> $p\nmid n$ <u>and</u> $p\nmid k$, <u>then</u> $\dim C=v$.

Proof. (1) Suppose that $p|n$. If $p|k$, then C is lf-orthogonal with respect to the ordinary dot product and have dim $C \leq \frac{1}{2}v$. If $p \nmid k$ similarly dim $C^{ext} \leq \frac{1}{2}(v+1)$. We ed only check that when $p \nmid k$, we have dim C^{ext}=dim C. The nal column of B is equal (mod p) to k^{-1} times the sum of e previous v columns. Discarding it, we next notice that e final row of the remaining matrix is equal to λk^{-1} times e sum of the first v rows. This demonstrates, as desired, at A and B have the same F_p-rank. Finally, the lower bound dim C is trivial.

(2) Suppose that $p \nmid n$ and $p|k$. Every row of A is thogonal (with respect to the ordinary dot product) to ,1,1,...,1). Thus dim $C \leq v-1$. The sum of all rows con- ining a 0 in the i-th column is the vector (n,...,n,0,n, ..,n) where the 0 is in the i-th column. These vectors nerate a (v-1)-dimensional subspace of $F_p{}^m$.

(3) Suppose that $p \nmid n$ and $p \nmid k$. Since |det A|= $\frac{1}{2}(v+1)$, the matrix A is invertible over F_p. Hence m C=v.□

The extended F_p-codes of projective planes are ther special in that the codewords of minimum weight in e code (and its dual) have geometric significance. The pic merits a lengthy digression.

The first v coodinates of the extended F_p-code of projective plane D correspond to the elements of P, the int set of D. Let us put the final coordinate on an equal oting by assigning it to a new point, which we label ∞. view the codewords then as functions from $P \cup \{\infty\}$ to F_p. e first v rows of the matrix B are the characteristic nctions of the sets $\ell \cup \{\infty\}$, where ℓ is a block. A useful ometric notion is that of an oval. An oval in a projective ane of even order n is a set S of n+2 points of P with the operty that any block meets S in at most 2 points. (Ovals arbitrary symmetric designs are explored in Problems 16 d 17.)

Theorem 2.7. Suppose that D is a projective plane order n and that p is a prime divisor of n. Let D be the

<u>extended</u> F_p<u>-code of</u> D <u>and let</u> D^ψ <u>be its dual with respect to the scalar product</u> ψ <u>defined above</u>.

<u>The minimum weight of</u> D^ψ <u>is</u> n+2. <u>Moreover, the codewords of weight</u> n+2 <u>are precisely the multiples of the characteristic functions of</u>

(1) <u>the extended blocks, and</u>

(2) <u>if</u> p=2, <u>the ovals</u>.

<u>Proof</u>. We know that $D \subseteq D^\psi$ and therefore that the multiples of the characteristic functions of the extended blocks are codewords of weight n+2 in D^ψ. Suppose that p=2. It is not hard to show that in a projective plane of even order a block meets an oval in 0 or 2 points. (See Problem 14.) The multiples of the characteristic function of the ovals are therefore orthogonal to the characteristic functions of the blocks and to $(\lambda, \ldots, \lambda, k)$. Hence the characteristic functions of the ovals are orthogonal to all vectors in D; that is, they lie in D^ψ.

We must now show that any vector u of nonzero weight $\leq n+2$ must be of one of these two types. We can conveniently view u as a function from $P \cup \{\infty\}$ to F_p (since the coordinates of u correspond to the elements of the set $P \cup \{\infty\}$). Let S be the set of $x \in P \cup \{\infty\}$ such that $u(x) \neq 0$.

<u>Case 1</u>: $\infty \in S$. Say $u(\infty) = \alpha$. Since u is orthogonal to all extended blocks, S must certainly contain further points. Let p and q be two points in $S \cap P$. We claim that u is α times the characteristic function of $\ell \cup \{\infty\}$, where ℓ is the unique block through p and q. Suppose to the contrary that some point y of $\ell \cup \{\infty\}$ does not lie in S. Let $\ell_1, \ldots, \ell_n$ be the n blocks (other than ℓ) through y. The characteristic function of $\ell_i \cup \{\infty\}$ is orthogonal to u. Since both are nonzero at ∞, the block ℓ_i must meet S in some point x_i (for $i=1, \ldots, n$). The points $\infty, p, q, x_1, \ldots, x_n$ are distinct (Why?), which means that u has weight at least n+3, contradicting our hypothesis. Hence, every point of $\ell \cup \{\infty\}$ lies in S. Since $|S| \leq n+2$, we have $S = \ell \cup \{\infty\}$. Finally we must show that $u(x) = \alpha$ for $x \in S$. Let $x \in S \cap P$ and let ℓ^* be a block through x other than ℓ. The scalar product of the characteristic

nction of $\ell^* \cup \{\infty\}$ and u equals $u(x)-u(\infty)$, which must be zero.
ence $u(x)=\alpha$ for all $x \in S$.

Case 2: $\infty \notin S$. Choose a point y in S. Because of
rthogonality, every extended block through y meets S in at
east one further point. This accounts for n+2 points in S.
ccordingly, every extended block through y meets S in
xactly one more point. Since the choice of y was arbitrary,
is an oval. Moreover, orthogonality forces $u(x)=-u(y)$ for
y distinct points x and y in S. This is impossible unless
=2, in which case u is the characteristic function of an
val.□

Remark. In the case p=2, the characteristic
nctions of the ovals lie in D^{ψ}. They will lie in D if
d only if $n \equiv 2 \pmod 4$. (See Problems 13, 14, 15.)

Among projective planes, PG(2,4) is particularly
emarkable. A 2-design, it is three times extendable to a
-(24,8,1) design admitting the 5-transitive Mathieu group
$_{24}$. While there are many proofs of this fact, by far the
ost elementary construction emerges from a study of the
inary code of the projective plane. We develop it below,
howing first how to extend PG(2,4) in three different (but
somorphic) ways to a 3-(22,6,1) design and then going on
o construct the 5-(24,8,1) design.

What are the ovals of PG(2,4)? If x_1, x_2, x_3, x_4 form
quadrilateral (that is, they are four points with no three
n a common block) in PG(2,4) and ℓ_{ij} is the block containing
$_i$ and x_j (for i,j=1,2,3,4 and i<j) there are exactly 2
oints x_5 and x_6 lying on no ℓ_{ij} and $\{x_1,x_2,x_3,x_4,x_5,x_6\}$ is
nd oval. So, every quadrilateral is contained in a
nique oval. The number of ovals must then be

$$\frac{21.20.16.9}{6.5.4.3} = 168.$$

urther counting shows that every triangle is contained in
ovals, every pair of points in 12 ovals, and every point
n 48 ovals. Moreover, any oval meets 40 ovals in exactly
points, 45 ovals in exactly 2 points, 72 ovals in exactly 1
oint and 10 ovals in no points.

Let D be the extended $\mathbb{F}_2$-code of PG(2,4). By brute force, we could compute that D has dimension 10. There is, however, a much simpler way. The code has length 22. Easy counting arguments (See Problems 6 and 9) show that dim $D \geq 10$. If the dimension were 11, then necessarily $D = D^{\psi}$. By Theorem 2.7, every oval would lie in D. However, in this case D would contain the characteristic functions of all ovals--contrary to the remark above. Hence dim D=10 and dim $D^{\psi}=12$.

There are exactly three codes E_1, E_2, E_3 of dimension 11 such that $D \subseteq E_i \subseteq D^{\psi}$ (for i=1,2,3). Let f be the characteristic function of an oval O and suppose that E_1 is the code generated by D together with f. Since f is orthogonal to itself and to D, the code E must be self-orthogonal (in fact, self-dual). Now, E_1 cannot contain the characteristic functions of the 112 ovals which meet O in 1 or 3 points. Thus E_1 contains the characteristic functions of at most 56 ovals. Similarly for E_2 and E_3. However, since each of the 168=56.3 ovals lies in one of these codes, each of the E_1, E_2 and E_3 must have exactly 56 characteristic functions of ovals. This naturally partitions the ovals into three sets of 56, called extension classes O_1, O_2, O_3. As E_i is self-orthogonal, the ovals in the extension class O_i must meet in 0 or 2 points (i=1,2,3). Ovals from different extension classes meet in an odd number of points.

Choose one of the extension classes, say O_1.

Proposition 2.8. *The incidence structure* G *whose* 22 *points are the elements of* $P \cup \{\infty\}$ *and whose* 77 *blocks are the* 21 *extended blocks of* PG(2,4) *and the* 56 *ovals in the extension class* O_1 *is a* 3-(22,6,1) *design*.

Proof. Because of the intersection properties, any three points of $P \cup \{\infty\}$ lie in at most one of the 77 blocks. These 77 blocks contain among them $77.\binom{6}{3} = 1540$ sets of three points. Since $1540 = \binom{22}{3}$, every set of three points is contained in a unique block.□

Let x and y be two points of the 3-(22,6,1) design G. The derived designs G_x and G_y are projective planes of order 4. By Problem 5, there is a unique projective plane of order 4, up to isomorphism. Thus G_x and G_y are isomorphic.

e can extend this to an automorphism of G to itself by ending x to y. Hence the automorphism group of G is transi- ive on points. Since $G_x \simeq PG(2,4)$ has a 2-transitive auto- orphism group, the full automorphism group of G is 3- ransitive. It is the Mathieu group M_{22}.

In order to construct the 5-(24,8,1) design, we ust study, in addition to the ovals, the subplanes of order in PG(2,4). A similar argument shows that any quadri- ateral of PG(2,4) lies in a unique subplane of order 2. here are therefore

$$\frac{21.20.16.9}{7.6.4.1} = 360$$

uch subplanes. Let π be such a subplane of order 2. Every lock of PG(2,4) meets π in 1 or 3 points. Accordingly, the haracteristic function of $\pi\cup\{\infty\}$, which we call an <u>extended</u> <u>ano plane</u>, lies in D^ψ. None of these vectors lie in D since, among other reasons, every Fano plane shares exactly points of P with 105 other planes and so the characteristic unctions of the corresponding extended Fano planes are not rthogonal). So, just as with the ovals, the Fano planes all into three classes F_1, F_2, F_3 corresponding to E_1, E_2, E_3. ano planes in the same class intersect in an odd number of oints of P; those in different classes meet in an even umber of points of P.

Moreover, if O is an oval in O_i and π is a plane n F_j then O and π share an even number of points of P if nd only if i=j.

We construct a 5-(24,8,1) design D_{24} as follows. he points are the 21 points of PG(2,4) together with three urther points called $\{\infty_1, \infty_2, \infty_3\}$. The blocks are the sets f the form

(a) $\ell\cup\{\infty_1,\infty_2,\infty_3\}$, where ℓ is a block of PG(2,4),
(b) $O\cup(\{\infty_1,\infty_2,\infty_3\} - \{\infty_i\})$ where O is an oval in extension class O_i (i=1,2,3),
(c) $\pi\cup\{\infty_i\}$, where π is a Fano plane in F_i (i=1,2,3)
(d) $\ell_1 \Delta \ell_2$, the symmetric difference of blocks of PG(2,4).

Theorem 2.9. The incidence structure D_{24} is a 5-(24,8,1) design.

Proof. By the properties of the extension classes any two blocks of D_{24} meet in an even number of points. Hence the binary code C_{24} of length 24 generated by the characteristic functions of the blocks of D_{24} is self-orthogonal with respect to the dot product.

We claim that every codeword has weight divisible by 4. Certainly this is true of our generating set for C_{24}, the blocks of D_{24}, since they all have weight 8. Let x and y be two codewords of weight divisible by 4. Then

$$\text{weight}(x+y)=\text{weight}(x)+\text{weight}(y)-2\,\text{weight}(x\cap y)$$

$$\equiv 0 \pmod 4,$$

since $\text{weight}(x\cap y)\equiv 0 \pmod 2$ by self-orthogonality. Thus every codeword of C_{24} has weight $\equiv 0 \pmod 4$.

Moreover C_{24} has no codeword of weight 4. (To see this requires some work. Check that no word of weight 4 can be orthogonal to every characteristic vector of the form $\ell\cup\{\infty_1,\infty_2,\infty_3\}$ with ℓ a block of PG(2,4).) Thus C_{24} has minimum weight 8.

Let b_1 and b_2 be distinct blocks of D_{24}. If $|b_1\cap b_2|\geq 5$ then the sum of the characteristic functions of b_1 and b_2 has weight at most 6, which is impossible. Hence every set of five points lies in at most one block. Each of the 759 blocks contains $\binom{8}{5}$ sets of five points. Since

$$759\binom{8}{5} = \binom{24}{5}$$

every set of five points lies in precisely one block.□

Remarks. (1) The code C_{24} is the celebrated extended Golay code of length 24.

(2) The 5-design D_{24} admits a 5-transitive group, namely the Mathieu group M_{24}.

Before leaving this topic, we observe that two symmetric designs can be found lurking in the structures above.

Proposition 2.10. Fix a block ℓ of PG(2,4). The 6 points of PG(2,4) off ℓ and the 16 ovals in any extension class which are disjoint from ℓ form the points and blocks of a symmetric (16,6,2) design.

Also, define an incidence structure S as follows. The points of S are the 56 ovals of an extension class. The blocks are the 56 sets consisting of an oval and the 10 ovals which are disjoint from it.

Proposition 2.11. The structure S is a symmetric (56,11,2) design.

The reader should verify the details of both propositions.

§2.3 THE MODULE OF A SYMMETRIC DESIGN

We now return to investigate more closely the extended F_p-code of a symmetric design. This code is not only self-orthogonal (with respect to ψ), it is sometimes self-dual.

Theorem 2.12. Let D be a symmetric (v,k,λ) design. Suppose that p is a prime such that $p\|n$ and $p\nmid\lambda$. Then the extended F_p-code of D has dimension $\frac{1}{2}(v+1)$ and is self-dual with respect to the scalar product $(\bar{x},\bar{y}) = x_1y_1 + \ldots + x_vy_v - \lambda x_{v+1}y_{v+1}$.

Proof. Since the extended F_p-code C^{ext} is self-orthogonal with respect to ψ we have $\dim C^{ext} \le \frac{1}{2}(v+1)$. The dimension is the F_p-rank of the extended incidence matrix B. According to the theory of invariant factors, there exist integral matrices P and Q with determinant 1 such that PBQ is a diagonal matrix. (Concerning invariant factors, see Appendix C). Say,

$$PBQ = \begin{pmatrix} d_1 & & 0 \\ & \ddots & \\ 0 & & d_{v+1} \end{pmatrix}$$

Since $|\det PBQ| = |\det B| = n^{\frac{1}{2}(v+1)}$ and $p\|n$, at most $\frac{1}{2}(v+1)$ of the d_i are multiples of p. Thus PBQ and hence B have F_p-rank at least $\frac{1}{2}(v+1)$. Hence $\dim C^{ext} = \frac{1}{2}(v+1)$.□

The fact that C^{ext} is self-dual has important ramifications. As noted in Appendix B, only certain $\mathbb{F}_p$-vector spaces with scalar products possess self-dual codes. Accordingly, Theorem 2.11 must entail some restrictions upon the possible parameters of a symmetric design. (The reader might wish to stop here and discover independently these restrictions.)

The proof above fails completely if $p^2|n$. In general the dimension of C^{ext} is much smaller than $\frac{1}{2}(v+1)$. The dimensions of the $\mathbb{F}_2$-codes of $PG(2,2^s)$, for example, have been computed exactly by algebraic coding theorists:

order of plane	[length, dimension] of code
2	[8,4]
4	[22,10]
8	[74,28]
16	[274,82]
.	.
.	.
.	.
2^s	$[2^{2s}+2^s+2, 3^s+1]$

The self-dual code appears to be the lucky exception, occurring only when $p \| n$. In some sense, though, this is not at all the case. We shall shortly see how other self-dual codes are lurking. If, emulating the proof of Theorem 2.12, we compute the invariant factors of the matrices B for $PG(2,2^s)$ we find:

order of plane	invariant factors
2	$(1,1,1,1,2,2,2,2)$
4	$(\overbrace{1,\ldots,1}^{10}, \overbrace{2,2}^{2}, \overbrace{4,\ldots,4}^{10})$
8	$(\overbrace{1,\ldots,1}^{28}, \overbrace{2,\ldots,2}^{9}, \overbrace{4,\ldots,4}^{9}, \overbrace{8,\ldots,8}^{28})$
16	$(\overbrace{1,\ldots,1}^{82}, \overbrace{2,\ldots 2}^{36}, \overbrace{4,\ldots,4}^{38}, \overbrace{8,\ldots,8}^{36}, \overbrace{16,\ldots,16}^{82})$

e symmetry is not coincidental. In the case when s is odd will provide us with a way to "slice off" a code of dimenon $\frac{1}{2}(v+1)$. To do this, we cannot work in characteristic but instead in characteristic zero. Accordingly we gress to discuss $\mathbb{Z}$-modules and lattices.

A <u>lattice</u> L of <u>rank</u> m is a subset of the rational ctor space Q^m which is a free $\mathbb{Z}$-module of rank m. In her words, L consists of all integral linear combinations some m linearly independent vectors. The lattices with ich we shall be concerned shall be those generated by the ws of some rational m x m matrices (such as incidence trices). In this case, we call the matrix a <u>generating trix</u> for the lattice. Notice that if V and U are gener-ing matrices for the same lattice L then V=PU for some tegral unimodular matrix P, and conversely. (A matrix is <u>imodular</u> if its determinant is a unit, i.e., ±1.)

There are many ways to obtain an F_p-code from a ttice L, for any prime p. If $L \subseteq \mathbb{Z}^m$, the simplest is to ke the reduction (mod p) of L. We can generalize this nstruction to obtain an infinite sequence of F_p-codes from as follows. Let $\pi^m: \mathbb{Z}^m \longrightarrow F_p{}^m$ be the "reduction (mod p)" momorphism. Define the F_p-codes:

$$C_\alpha = \pi^m(p^{-\alpha}L \cap \mathbb{Z}^m)$$

r $\alpha = \ldots,-2,-1,0,1,2,\ldots$. Clearly we have

$$\ldots \subseteq C_{-1} \subseteq C_0 \subseteq C_1 \subseteq C_2 \subseteq \ldots .$$

e dimension of the code C_α is related to the invariant ctors of a generating matrix C for L. For simplicity's ke, we assume that $L \subseteq \mathbb{Z}^m$ (if not, we can first multiply e lattice by a suitable constant scalar).

<u>Proposition 2.13</u>. <u>Let</u> C <u>be an integral matrix</u> <u>ich generates the lattice</u> $L \subseteq \mathbb{Z}^m$ <u>and let</u> π_i <u>be the number</u> <u>f its invariant factors</u> d <u>such that</u> $p^i \| d$. <u>Then</u> dim $C_\alpha = 0$ <u>r</u> $\alpha<0$ <u>and</u>

$$\dim C_\alpha = \pi_0 + \ldots + \pi_\alpha \quad \underline{\text{for}}\ \alpha \geq 0.$$

Proof. Since $L \subseteq \mathbb{Z}^m$, it is clear from the construction that $\dim C_\alpha = 0$ for $\alpha < 0$. Now let P and Q be integral unimodular matrices such that PCQ is a diagonal matrix D. Say

$$PCQ = D = \begin{pmatrix} d_1 & & 0 \\ & \ddots & \\ 0 & & d_m \end{pmatrix}.$$

So, $PC = DQ^{-1}$ and then DQ^{-1} is an alternative generating matrix for L. Now the rows of DQ^{-1} are vectors $d_1v_1, \ldots, d_mv_m$ for some vectors $v_1, \ldots, v_m \in \mathbb{Z}^m$. Thinking about the construction, C_α will have as a basis the set $\{\pi^m(v_i) \mid d_i \neq 0 \pmod{p^{\alpha+1}}\}$. Hence $\dim C_\alpha = \pi_o + \ldots + \pi_\alpha$.□

Next, we introduce the notion of the dual of a lattice. Suppose that ϕ is a nondegenerate scalar product defined on Q^m. The dual of L with respect to ϕ, denoted $L^{[\phi]}$, is the set

$$L^{[\phi]} = \{x \in Q^m \mid \phi(x,y) \in \mathbb{Z} \quad \text{for all } y \in L\}.$$

$L^{[\phi]}$ is also a lattice. In fact if L has a basis $\{v_1, \ldots v_m\}$ then a basis for $L^{[\phi]}$ is $\{v_1^*, \ldots, v_m^*\}$--i.e., the dual basis satisfying

$$\phi(v_i, v_j^*) = \begin{cases} 1 & \text{if } i=j, \\ 0 & \text{otherwise.} \end{cases}$$

In terms of matrices, if U is a generating matrix for L then the matrix V such $U\phi V^T = I$ will surely be a generating matrix for $L^{[\phi]}$ (where we let ϕ also denote the matrix of the bilinear form). So $V = (U^{-1})^T(\phi^{-1})^T$ is a generating matrix for $L^{[\phi]}$.

For certain lattices, the dual happens to be just a multiple of the lattice. A lattice is called a-modular (for a rational number a) if

$$L^{[\phi]} = aL,$$

ere $aL = \{ax \mid x \in L\}$. Notice that if L is a-modular then for
l x, $y \in L$ we have $\phi(x,y) = a^{-1}s$ for some integer s.

If a lattice is a-modular, it turns out that the variant factors of a generating matrix display just the rt of symmetry we observed above. In the sequel, fix a ime p, an integer s, and a nonsingular scalar product ϕ ch that $(\mathbb{Z}^m)^{[\phi]} = (\mathbb{Z}^m)$ or, equivalently, such that the trix ϕ is integral and unimodular.

Proposition 2.14. Let C be an integral matrix ich generates the lattice $L \subseteq \mathbb{Z}^m$ and let π_i be the number its invariant factors d such that $p^i \| d$.

Suppose that L is p^{-s}-modular with respect to ϕ. en

$$\pi_i = \pi_{s-i}$$

d

$$\pi_i = 0$$

r $i<0$ and $i>s$.

Proof. We have $L = p^s L^{[\phi]}$. Since $L^{[\phi]}$ is generated $(C^{-1})^T(\phi^{-1})^T$ then L is generated by $C^* = p^s(C^{-1})^T(\phi^{-1})^T$. course, L is also generated by C. Thus C and C^* have the me invariant factors. The number of invariant factors d C^* such that $p^i \| d$ is π_{s-i}. Hence $\pi_i = \pi_{s-i}$.

Since C is integral $\pi_i = 0$ for $i<0$ and thus also $= 0$ for $i>s$.□

This symmetry has important implications for the ain of codes C_α defined above. Note that since we have sumed that ϕ is integral and unimodular, then ϕ induces a ll-defined inner product ϕ on $\mathbb{F}_p{}^m = \pi^m(\mathbb{Z}^m)$. (Why?)

Theorem 2.15. Suppose that L is a lattice such at $L \subseteq \mathbb{Z}^m$ and L is p^{-s}-modular with respect to ϕ. Let

$$C_\alpha = \pi^m(p^{-\alpha}L \cap \mathbb{Z}^m)$$

for all integers α. Then

$$\{0\} = C_{-1} \subseteq C_0 \subseteq \dots \subseteq C_{s-1} \subseteq C_s = \mathbb{F}_p^{\,m}.$$

Moreover,

$$(C_\alpha)^{\overline{\phi}} = C_{(s-1)-\alpha}.$$

In particular, if s is odd, $C_{\frac{1}{2}(s-1)}$ is a self-dual code with respect to $\overline{\phi}$.

Proof. We have already verified the chain of inclusions and the fact that $C_{-1} = \{0\}$. Now, because L is p^{-s}-modular, then $p^{-s}\phi(x,y)$ is an integer for all $x, y \in L$. Thus

$$\phi(p^{-\alpha}x,\ p^{-((s-1)-\alpha)}y) \equiv 0 \pmod p,$$

whence we see that

$$C_{(s-1)-\alpha} \subseteq (C_\alpha)^{\overline{\phi}}.$$

To prove that they are equal, we must verify that $\dim(C_{(s-1)-\alpha}) + \dim(C_\alpha) = m$. Now, by Proposition 2.14,

$$\begin{aligned}\dim(C_{(s-1)-\alpha})+\dim(C_\alpha) &= (\pi_0+\dots+\pi_{(s-1)-\alpha})+(\pi_\alpha+\dots+\pi_0)\\ &= (\pi_0+\dots+\pi_{(s-1)-\alpha})+(\pi_{s-\alpha}+\dots+\pi_s)\\ &= m.\end{aligned}$$

Last of all, we observe that $C_s = (C_0)^{\overline{\phi}} = (\{0\})^{\overline{\phi}} = \mathbb{F}_p^{\,m}$, which completes the proof.□

We see now how to produce self-dual codes over $\mathbb{F}_p$ from p^{-s}-modular lattices L (when s is odd). The only question is how to recognize when a given generating matrix will generate a p^{-s}-modular lattice (with respect to a particular inner product). This, of course, we already know how to do.

Proposition 2.16. Let C be a rational matrix and t ϕ be a scalar product on Q^m (with its matrix also denoted ϕ). Then the lattice L generated by the rows of C is ^{s}-modular if and only if

$$C\phi C^T = p^s U$$

r some integral unimodular matrix U.

Now let us apply our results to symmetric designs. fine the $\mathbb{Z}$-module of a symmetric (v,k,λ) design D to be e lattice M generated by the rows of the extended incidence trix

$$B = \left(\begin{array}{c|c} A & \begin{matrix}1\\ \vdots\\ 1\end{matrix} \\ \hline \lambda \ldots \lambda & k \end{array}\right).$$

shall use the scalar product ψ, whose matrix is

$$\psi = \begin{pmatrix} 1 & & & 0 \\ & \ddots & & \\ & & 1 & \\ 0 & & & -\lambda \end{pmatrix}.$$

can exploit the equation $B\psi B^T = n\psi$, which we observed in .1.

Suppose for example that D is PG$(2,2^s)$. Then ψ is integral unimodular matrix and

$$B\psi B^T = 2^s\psi.$$

nce the $\mathbb{Z}$-module of PG$(2,2^s)$ is 2^{-s}-modular with respect ψ, by Proposition 2.16. This explains the symmetry ong the invariant factors. Moreover, by applying Theorem 15, we obtain a chain of $\mathbb{F}_2$-codes. Whenever s is odd, e code $C_{\frac{1}{2}(s-1)}$ is self-dual with respect to the ordinary t product.

We should like to apply the same method to an arbitrary symmetric (v,k,λ) design with $p^s \| n$, to produce a chain of $\mathbb{F}_p$-codes, including a code self-dual with respect to an appropriate scalar product whenever s is odd (or, equivalently, whenever p divides n*, the square-free part of n). Two minor problems arise. First, n need not be simply a power of p, and second, ϕ will not in general be unimodular. With some slight modifications in the method we can remedy these problems.

Primes other than p really do not enter into the construction. The best way to prevent them from being a nuisance is to declare them to be units! That is, we should perform the above construction using not $\mathbb{Z}$-modules but modules over $\mathbb{Z}_{(p)}$, the integers localised at the prime p. Over this ring, all primes other than p are units. All the proofs above go through as before. So, in order to use the equation $B\psi B^T = n\psi$ to produce an appropriate chain of codes we in fact only require that $|\det\psi| = \lambda$ not be divisible by p.

Finally, if $|\det\psi|$ happens to be divisible by p, we must slightly modify the matrices B and ψ in order for the construction to work and for us to obtain $\mathbb{F}_p$-codes self-dual with respect to an appropriate scalar product. We make this precise in the proof of the following result.

Theorem 2.17. Suppose that D *is a symmetric* (v,k,λ) *design and that* p *is a prime such that* p *divides* n*, *the square-free part of* n. *Then*

(1) *If* $p \nmid \lambda^*$, *we can associate to* D *an* $\mathbb{F}_p$*-code of length* v+1 *which is self-dual with respect to the non-degenerate scalar product*

$$\psi(\bar{x},\bar{y}) = x_1y_1+\ldots+x_vy_v - \lambda^* x_{v+1}y_{v+1}.$$

(2) *If* $p | \lambda^*$, *we can associate to* D *an* $\mathbb{F}_p$*-code of length* v+1 *which is self-dual with respect to the non-degenerate scalar product*.

$$\psi(\overline{x},\overline{y}) = x_1y_1+\ldots+x_vy_v + (\frac{\lambda^* n^*}{p^2})x_{v+1}y_{v+1}.$$

<u>Proof</u>. Since $p|n^*$, say that $p^s \| n$ with s odd. ıppose first that $p \nmid \lambda^*$. Write $\lambda=t^2\lambda^*$. Let

$$B_1 = \left(\begin{array}{c|c} A & \begin{matrix} t \\ \vdots \\ t \end{matrix} \\ \hline \lambda\ldots\lambda & tk \end{array}\right) \quad \text{and} \quad \psi_1 = \begin{pmatrix} 1 & & & 0 \\ & \ddots & & \\ & & 1 & \\ 0 & & & -\lambda^* \end{pmatrix} .$$

ıen $B_1\psi_1B_1^T = n\psi_1$. Now ψ_1 is a unimodular matrix over $\mathbb{Z}_{(p)}$, ıe integers localised at p. So, we may apply our method to ıtain a chain of codes C_α. The code $C_{\frac{1}{2}(s-1)}$ is self-dual .th respect to ψ_1.

Next, suppose that $p|\lambda^*$. In view of the identity .' = n(n-1), where λ' is the parameter of the complementary ›sign, we have $p \nmid (\lambda')^*$. So, by the previous paragraph, we ıy use the complementary design to generate a chain of)-codes, including an F_p-code self-dual with respect to the ›ndegenerate scalar product

$$\psi'(\overline{x},\overline{y}) = x_1y_1+\ldots+x_vy_v + (\lambda')^*\, x_{v+1}y_{v+1}.$$

ince $\lambda\lambda' = n(n-1)$, we have $\lambda^2\lambda' = (\lambda n)(n-1)$. Hence $(\lambda')^* \equiv$ $^2(-\lambda^*n^*/p^2)$ (mod p), for some integer f. Thus, if we ıltiply the final coordinate of our codes by f, they will ıtisfy duality properties with respect to

$$\psi(\overline{x},\overline{y}) = x_1y_1+\ldots+x_vy_v - (\lambda^*n^*/p^2)x_{v+1}y_{v+1}$$

the advantage of this latter form being that it involves ıe parameters of D rather than D'). □

As we mentioned earlier, not all vector spaces with calar products possess self-dual codes. In Appendix C, we ıow that a vector space over F_p (with p odd) of dimension v+1) possesses a code which is self-dual with respect to a nondegenerate scalar product ψ if and only if $(-1)^{\frac{1}{2}(v+1)}\det\psi$ is a square in F_p. This immediately proves the following consequence of Theorem 2.17.

Theorem 2.18. Suppose that there exists a symmetric (v,k,λ) design. Then for every odd prime p,

(i) If $p|n^*$ and $p\nmid\lambda^*$, then $(-1)^{\frac{1}{2}(v-1)}\lambda^*$ is a square (mod p).

(ii) If $p|n^*$ and $p|\lambda^*$, then $(-1)^{\frac{1}{2}(v+1)}(\lambda^* n^*/p^2)$ is a square (mod p).

Of course, Theorem 2.18 is precisely the second and third conditions in Theorem 2.5, our alternate form of the Bruck-Ryser-Chowla Theorem. Thus part of the Bruck-Ryser-Chowla Theorem can be thought of as requiring the existence of certain self-dual codes which would be constructible from the putative symmetric (v,k,λ) design.

What about the first condition of Theorem 2.5? It is worth noting that when $p\nmid n$ the assertion is trivial since $\lambda\lambda' = n(n-1)$ implies that $n\equiv 1 \pmod p$, so n is certainly a square (mod p). Thus for designs with $(k,\lambda) = 1$ we may ignore this condition entirely. In this case, the B-R-C conditions are precisely equivalent to the existence of appropriate self-dual codes. When $p|n$, the first condition of Theorem 2.5 provides a nontrivial restriction on the parameters. I cannot see any elementary way to give this condition a coding theoretic interpretation. Perhaps the reader can supply such an argument, yielding a complete proof of the B-R-C Theorem by coding theory.

In any case, the purpose of this discussion is not primarily to provide a new proof of an already known result. Rather the chain of codes constructed above will provide the point of departure for further investigation in subsequent chapters.

PROBLEMS - CHAPTER 2

. Show, by using Theorem 2.5, that if a projective plane of rder n exists with $n \equiv 1$ or 2 (mod 4), then any prime dividing he square-free part of n is congruent to 1 (mod 4). This ondition is equivalent to the condition that n is the sum f two integral squares. (Prove this by number theory or ee [56].)

. What does the Bruck-Ryser-Chowla Theorem say about the ossible parameters of a symmetric (v,k,2) design? a adamard design?

The next two problems concern certain number-theoretic onsequences of the relations between the parameters v,k,λ nd n.

. Suppose that D is a symmetric (v,k,λ) design in which v s a power of 2. We prove here that v=4n and hence that the arameters are

$$(v,k,\lambda) = (2^{2m}, 2^{2m-1} \pm 2^{m-1}, 2^{2m-2} \pm 2^{m-1})$$

$$\text{and} \quad n = 2^{2m-2}$$

or some integer m.

To begin, let $v=2^e$. After possibly replacing D by ts complement assume that $k<\frac{1}{2}v$. Since v is even, write $=m^2$ and let f be the integer such that $2^f \| m$. Consider two ases: $2f \geq e-2$ and $2f \leq e-2$.

(i) If $2f \geq e-2$, conclude from the inequality $n<\frac{1}{2}v$ hat $2f=e-2$ and $v=4n$.

(ii) If $2f \leq e-2$, use the identity $k^2=n+v\lambda$ to show hat 2^f divides k and λ. Next use $v\lambda = k^2-n = (k+m)(k-m)$ to prove that 2^{e-1} divides either k+m or k-m. By estimating the sizes of k+m and k-m conclude that $k+m = 2^{e-1}$. Finally, use $k^2 - v\lambda = n = (2^{e-1}-k)^2$ to show that v=4n and 2f=e-2.

4. Show that the previous problem characterizes the prime 2 in the following sense: if v and n are both powers of a prime p, then p=2.

(i) Suppose that $p \neq 2$. Let $v=p^e$ and $n=p^s$. Use the identity $n+v\lambda=(n+\lambda)^2$ to prove that s must be even and that $p^{\frac{1}{2}s}$ divides k and λ.

(ii) Set $m=\sqrt{n}=p^{\frac{1}{2}s}$. Arguing as in part (ii) of the preceding problem, show that p^e divides k+m or k-m. Reach a contradiction.

5. Show that there is exactly one symmetric (21,5,1) design up to isomorphism, namely PG(2,4).

(i) Review the discussion of ovals in a projective plane of order 4 and notice that we never used the fact that the plane is PG(2,4). The properties apply to any projective plane of order 4.

(ii) Let P be the set of points of a projective plane of order 4 and let H be an oval. Show that, to each point p of P-H, there corresponds a partition of H into three pairs (determined by the blocks through p) and that every such partition occurs.

(iii) Let D_1 and D_2 be projective planes of order 4 with ovals H_1 and H_2 respectively. Show that any one-to-one mapping of the points of H_1 to those of H_2 can be uniquely extended to an isomorphism of the planes. (Incidentally, we use this to find the order of PGL(3,4).)

The next five problems concern lower bounds for the dimension of the F_p-code of a symmetric (v,k,λ) design.

6. Show that the F_p-code of a projective plane of order n has dimension at least 3n-2. (Hint: Let $p_1,\ldots,p_{n+1}$ be the points of some block. Let ℓ_1 be the block through p_1 and p_2,

let $\ell_2,\ldots,\ell_{n+1}$ be the remaining blocks through p_1 and let $\ell_{n+2},\ldots,\ell_{2n}$ be all but one of the remaining blocks through p_2. Let $\ell_{2n+(i-2)}$ be a block through p_i for $i=3,\ldots,n$. Show that the characteristic functions of $\ell_1,\ldots,\ell_{3n-2}$ are linearly independent.) Hence show that the F_2-code of PG(2,4) has dimension at least 10.

7. Let D be a symmetric (v,k,λ) design and let A be its incidence matrix.

(i) Show that any d rows of A are linearly independent over any field, provided that $d<1+(k/\lambda)$.

(ii) Let C be the F_p-code of D, for some prime p. Show that a codeword of C can be expressed in at most one way as a linear combination of e or fewer rows, where $e=[\frac{1}{2}d]$. Show that dim C is at least $\log_p(1+(p-1)\binom{v}{1}+(p-1)^2\binom{v}{2}+\ldots+(p-1)^e\binom{v}{e})$.

8. An immediate consequence of the previous problem is that dim $C\geq\log_p(1+(p-1)v)$, for any symmetric design. Show that equality holds if and only if $p=2$ and D is isomorphic to the complement of $PG(2,2^s)$, for some s.

(i) Let B_1 and B_2 be two distinct blocks of the complement of $PG(2,2^s)$. Show that the symmetric difference $B_1\Delta B_2$ is also a block. Hence show that the F_2-code of this design has dimension s+1, which attains the bound $\log_2(1+v)$.

(ii) Now, suppose that D is a symmetric (v,k,λ) design whose F_p-code has dimension t and that $t=\log_p(1+(p-1)v)$. Identify all codewords of C and show that p must be 2. Thus $v=2^t-1$. Consider any t linearly independent columns of the incidence matrix A of D and show that every nonzero t-tuple of ones and zeroes occurs exactly once. The columns of A must be exactly the 2^t-1 nonzero linear combinations of these t columns. Thus, show that A is uniquely determined up to permutations of the rows and columns. Use (i) to complete the proof.

9. Consider the F_2-code of PG(2,4). The bound of Problem 6 states that dim $C\geq 10$, while that of Problem 7 yields only

dim $C \geq 8$. We can refine the basic idea behind Problem 7 to give an alternate proof that dim $C \geq 10$.

(i) By Problem 7, any five rows of the incidence matrix A of PG(2,4) are linearly independent. Show that six rows are linearly dependent if and only if the corresponding six blocks form an oval in the dual design.

(ii) Show that a codeword can be expressed in at most one of the following forms: (a) as a sum of 0,1 or 2 rows of A, (b) as a sum of three rows of A corresponding to blocks sharing a common point, (c) as a sum of three rows of A corresponding to blocks not sharing a common point. Show that in either of the first two cases the expression is unique, while in the last case a codeword can be expressed as such a sum in at most four ways. (Hint: three blocks sharing no common point lie in three ovals of the dual design.)

(iii) Hence show that C has at least

$$1 + \binom{21}{1} + \binom{21}{2} + 210 + \tfrac{1}{4}\left[\binom{21}{3} - 210\right] = 722$$

codewords. Hence dim $C \geq 10$.

10. Let D be a symmetric (v,k,2) design and let C be its F_p-code, for some prime p. Let η be the homomorphism which restricts each codeword to the k coordinates of some fixed block B.

(i) Considering separately the cases p=2 and p odd, find the dimension of the image of η.

(ii) If p=2, show that the kernel of η is not zero. (Hint: let x,y,z be three points of B. Consider the sum of the three blocks passing through exactly two of these points.)

(iii) Show that dim $C \geq k$.

(iv) Show that, in the case of a symmetric (16,6,2) design, this provides a better bound than Problem 6.

11. What further lower bounds can you devise for the dimension of the F_p-code of a symmetric design?

12. Verify the counting arguments used in the discussion of PG(2,4) in §2.2.

Let E be a projective plane of even order n and let D be the extended F_2-code of E. The next three problems concern whether the characteristic function of the ovals of E lie in D.

13. If $n \equiv 2 \pmod 4$, prove in two ways that the ovals lie in D--indirectly, by showing that $D=D^{\psi}$, and directly, by expressing the characteristic function of an oval as a sum of extended blocks.

14. Let C be a self-orthogonal code over F_2 generated by a set of codewords having weight divisible by 4. Prove that every codeword of C has weight divisible by 4.

15. If $n \equiv 0 \pmod 4$, prove that D does not contain the characteristic function of any oval. (Hint: Consider the subcode D' of D consisting of all codewords having a zero in the final coordinate. Show that D' is generated by the complements of the extended blocks. Apply the previous problem.)

16. An <u>arc</u> in a symmetric (v,k,λ) design is a set S of points such that no block contains more than two points of S. Call a block an <u>exterior</u>, <u>tangent</u>, or <u>secant</u> block to S according as it meets S in 0,1 or 2 points.

(i) Suppose that S has a tangent block which meets it at the point p. By counting pairs (q,B) where $p,q \in S \cap B$ and $p \neq q$, show that $|S| \leq (k+\lambda-1)/\lambda$.

(ii) Suppose that S has no tangent blocks. Show, in a similar manner, that $|S| = (k+\lambda)/\lambda$. Hence λ must divide k. Furthermore, by considering the blocks through a point not in S show that $(k+\lambda)$ is even. Hence, n is even.

Because of these results, Assmus and Van Lint call a set S of points an <u>oval</u> if it is as large an arc as one can reasonably hope for--that is, if

$$S = (k+\lambda-1)/\lambda \qquad \text{when either n is odd or } \lambda \nmid k$$

$$\text{or} \quad S = (k+\lambda)/\lambda \qquad \text{when n is even and } \lambda \mid k.$$

Show that this agrees with the definition already given for ovals in a projective plane of even order and that ovals in such planes have no tangents.

17. Consider the symmetric (11,5,2) design H(11).

(i) Show that any 3 points not in a block form an oval. Hence any 3 points lie on 2 blocks and 3 ovals. Any 3 points lie in either one block or form an oval.

(ii) Every oval has three tangents.

(iii) Define a new incidence structure D' as follows. The points of D' are the points and blocks of H(11). Blocks of D' are the sets of size 6 of the following type:

(a) a point of H(11) and the 5 blocks containing it,

(b) a block of H(11) and the 5 points contained in it,

(c) the three points of an oval and its three tangent blocks.

Show that D' is a 3-(22,6,1) design, i.e., an extension of PG(2,4).

18. What bounds can you give for the minimum weight of the $\mathbb{F}_p$-code of a symmetric (v,k,λ) design?

19. If L_1 and L_2 are lattices in $\mathbb{Q}^m$, then $L_1 + L_2 = \{x+y \mid x \in L_1 \text{ and } y \in L_2\}$ is a lattice. Suppose that we only ask that L_1 and L_2 be free $\mathbb{Z}$-modules of rank m in $\mathbb{R}^m$ (the real vector space). Must $L_1 + L_2$ be a free $\mathbb{Z}$-module of rank m in $\mathbb{R}^m$?

20. If L_1 and L_2 are lattices in Q^m and ϕ is a nonsingular scalar product, then

(i) $(L_1^{[\phi]})^{[\phi]} = L_1$,

(ii) $(L_1+L_2)^{[\phi]} = L_1^{[\phi]} \cap L_2^{[\phi]}$,

(iii) $(L_1 \cap L_2)^{[\phi]} = L_1^{[\phi]} + L_2^{[\phi]}$.

In the next two problems we compute the chain of codes associated with the symmetric design PG(m,2). The answer turns out to be a well-known and important class of codes.

21. Consider the vector space $V = \mathbb{F}_2^m$. For each i=0,...,m, we define a big matrix R_i whose columns are indexed by the $N=2^m$ elements of V (let the zero vector correspond to the last coordinate) and whose rows are the characteristic functions of the subspaces of dimension $\geq m-i$. The $\mathbb{F}_2$-span of R_i is called the i-th <u>order Reed-Muller code</u>, denoted RM(i,m).

(i) The matrix R_0 has only one row of all ones. So dim RM(0,m)=1.

(ii) Show that R_1 looks like

$$\begin{pmatrix} 1 & \cdots & 1 \\ & & \vdots \\ & A & \vdots \\ & & 1 \end{pmatrix}$$

where A is the incidence matrix of PG(m,2). Then RM(1,m) is just the extended code of PG(m,2). Reasoning as in Problem 8, show that it has dimension m+1. Let $\{\bar{1}, v_1, \ldots, v_m\}$ be a basis for it, where the v_i are characteristic functions for certain (m-1)-dimensional subspaces and $\bar{1}$ is the all-one vector.

(iii) If x and y are the characteristic functions of sets S and T, respectively, then their componentwise product, which we write xy, is the characteristic function of $S \cap T$. Now, any (m-2)-dimensional subspace of V is the intersection of two (m-1)-dimensional subspaces. So, show that RM(2,m) consists of all products of codewords in RM(1,m). Thus $\{\overline{1}, v_1, \ldots, v_m, v_1v_2, v_1v_3, \ldots, v_{m-1}v_m\}$ is a basis and RM(2,m) has dimension $1 + \binom{m}{1} + \binom{m}{2}$.

(iv) Show that dim RM(i,m) $= 1 + \binom{m}{1} + \binom{m}{2} + \ldots + \binom{m}{i}$.

22. Consider the extended incidence matrix of PG(m,2)

$$B = \left(\begin{array}{ccc|c} \lambda & \ldots & \lambda & k \\ \hline & & & 1 \\ & & & \cdot \\ & A & & \cdot \\ & & & \cdot \\ & & & 1 \end{array}\right).$$

Let M be the $\mathbb{Z}$-linear span of the rows of B. We defined C_i, the i-th code in the chain, to be the reduction (mod 2) of $p^{-i}M \cap \mathbb{Z}^{v+1}$. Show that $RM(i+1,m) \subseteq C_i$ for i=1,...,m-2. (Hint: let U be a subspace of dimension (m-i) and let x be the sum of the characteristic functions of all (m-1)-dimensional subspaces not containing U. Show that the entries of $2^{-i}x$ are integers, odd precisely in coordinates corresponding to points in V-U.)

Next, consider the inclusions:

$$RM(i+1,m) \subseteq C_i = C_{(s-1)-i}{}^{[\psi]} \subseteq RM(s-i,m)^{[\psi]}.$$

By computing the dimension of the two Reed-Muller codes show that equality holds throughout. Hence $RM(i+1,m) = C_i$ for i=0,1,...,m-2.

23. Suppose that there exists an integral v x v matrix C such that $CC^T = mI$. Show that if $v \equiv 2 \pmod 4$ then m must be the sum of two integral squares. (Use the method of §2.3.)

24. Must the codes in the chain defined in §2.3 be distinct? (Hint: Use the fact that dim C=16, where C is the $\mathbb{F}_2$-code

f H(31). If you don't wish to verify this by hand or lectronic computation, you may wait until Chapter 4, in hich it will follow from a more general result.)

NOTES TO CHAPTER 2

§2.1 The Bruck-Ryser-Chowla Theorem was proven in the case $\lambda=1$ by Bruck and Ryser [20] and in general by Chowla and Ryser [29]. The original proof used the Hasse-Minkowski theorem about the equivalence of rational quadratic forms, but more elementary proofs were discovered later. The proof in the text is a variation, due to the author, of the usual elementary proof. (See Ryser [115].) See Hardy and Wright [56], for Lagrange's Four-Squares Identity and Four-Square Theorem.

The first seventeen admissible parameters triples (v,k,λ)--ordered by v--are (7,3,1),(11,5,2),(13,4,1),(15,7,3), (16,6,2),(19,9,4),(21,5,1),(23,11,5),(25,9,3),(27,13,6), (31,6,1),(31,10,3),(31,15,7),(37,9,2),(40,13,4),(41,16,4), and (45,12,3). Designs with these parameters are known to exist. In fact, all but six can be realized as PG(m,q) or H(q). Three of the remaining six can be found here in: (16,6,2) in Example 4 of Chapter 1, (37,9,2) in Example 7 of Chapter 4, and (45,12,3) in Example 14(1) of Chapter 1 (with $\lambda=3$ and d=1). As for the remaining three, see [49] for (25,9,3), [12] for (31,10,3) and [51] for (41,16,6). The first open case is (49,16,5).

Constructions of projective planes have been given by many authors. See generally, Hughes and Piper [65] and Lüneburg [85].

Examples of designs with $v>\lambda^2(\lambda+2)$ can be found in: Aschbacher [10] for (79,13,2); Haemers [46] for (71,15,3); Example 7 of Chapter 4 for (37,9,2); Proposition 2.11 for (56,11,2).

§2.2 The theory of codes is a vast topic of great applicability. See generally, MacWilliams and Sloane [87]. Proposition 2.6 is often attributed to Hamada [54]. For relations between coding theory and particular symmetric designs, see MacWilliams, Sloane and Thompson [88], Hall [51], Cameron and van Lint [25], Salwach [117] and Assmus, Mezzaroba and Salwach [7].

§2.3 Assmus reports that he and Mattson first worked out the dimension of the binary code of a projective plane of order 10 by "direct but bare-handed" methods and that the proof of Theorem 2.12 in the general case is due to Gleason. The result became a folk theorem and, to my knowledge, first appeared in print in MacWilliams, Sloane and Thompson [88].

The dimensions of the F_p-codes of PG(m,q) have been worked out by Goethals and Delsarte [41], Graham and MacWilliams [44], MacWilliams and Mann [86] and Smith [127].

The main concepts of this section are due to Lander [78,79]. Lander also has determined the codes in the chain associated with PG(m,q) (by identifying them as particular cyclic codes) and has calculated their dimensions, thus generalizing the known formulae for the dimension of the usual F_p-codes of projective geometries. See [78]. These codes generalize the Reed-Muller codes.

Problems. Problems 3 and 4 are due to Mann [92]. Problems 6, 7, 9 and 10 are due to Lander. Problem 8 is essentially due to Hamada and Ohmori [55]. Problem 16 is taken from Assmus and van Lint's paper [4] on ovals in symmetric designs. Problem 17 is an example of a fascinating connection between projective planes and biplanes (symmetric designs with $\lambda=2$). See [7], [8] and the discussion in [25, p. 92-93]. For more on Reed-Muller codes, which are very important in coding, see [87]. Problem 21 gives another proof of a result by Delsarte, Goethals and Seidel [31].

3. AUTOMORPHISMS

3.1 FIXED POINTS AND BLOCKS

An automorphism group of a symmetric (v,k,λ) design can be viewed as a permutation group on a set of v objects in two ways--corresponding to its action on points or its action on blocks. These permutation representations are in general different. Nevertheless, they are intimately related.

Theorem 3.1. An automorphism σ of a symmetric design fixes an equal number of points and blocks.

Proof. If A is the incidence matrix of the symmetric design, σ specifies permutation matrices P and Q such that

$$PAQ = A,$$

where P acts to permute blocks and Q acts to permute points. The number of objects fixed by a permutation is precisely the trace of the associated permutation matrix. Since A is non-singular and since the inverse of a permutation matrix is simply its transpose, then

$$AQA^{-1} = P^T.$$

Thus Q and P^T are similar and trace Q = trace P.□

For a matrix-free proof, see Problem 1.

Corollary 3.2. An automorphism σ of a symmetric design has the same cycle structure, whether considered as a permutation on points or on blocks.

Proof. Let π_1 be the point-permutation and let π_2 be the block-permutation corresponding to σ. Let $f_i(d)$ be the number of cycles of length d in the cycle decomposition of π_i (for i=1,2). For every integer m, the number of fixed points of π_i^m is

$$\sum_{d|m} df_i(d).$$

By Theorem 3.1,

$$\sum_{d|m} df_1(d) = \sum_{d|m} df_2(d)$$

for all integers m. By a simple induction (or an application of the Mobius inversion formula), $f_1(d) = f_2(d)$ for all d.□

Theorem 3.1 definitely does not mean that any automorphism group of a symmetric design fixes as many points as blocks. The subgroup stabilizing a point in the full automorphism group of PG(2,2) provides a counter-example: it fixes one point and no blocks. The following, however, is true.

Theorem 3.3. An automorphism group G of a symmetric design has as many orbits on points as on blocks.

Proof. By the Cauchy-Frobenius Lemma (see Appendix A), the number t of orbits of G is given by

$$t = \frac{1}{|G|} \sum_{g \in G} |Fix(g)|,$$

where Fix(g) is the set of objects fixed by G. The result now follows from Theorem 3.1.□

An automorphism group is therefore transitive on points if and only if it is transitive on blocks. It is regular on points if and only if it is regular on blocks. There is no ambiguity then in saying simply that an automorphism group is transitive, or regular.

The rank r of a transitive permutation group G on a set X is defined (in Appendix A) to be the number of orbits of G_x, the stabilizer of the object x, on X. (Equivalently it is the number of orbits of G on X x X.) By Proposition A.3, the rank is given by

$$r = \frac{1}{|G|} \sum_{g \in G} |Fix(g)|^2,$$

which shows that the rank is completely determined by the number of fixed points of each $g \in G$. Hence,

Theorem 3.4. A transitive automorphism group of a symmetric design has the same rank whether considered as a permutation group on points or on blocks.

A transitive permutation has rank 2 if and only if it is 2-transitive. So, again, an automorphism group is 2-transitive on points if and only if it is 2-transitive on blocks.

All of these results depend essentially on the fact that an automorphism fixes an equal number of points and blocks. It is interesting (and more complex) to ask about how the fixed points and fixed blocks are interrelated geometrically. Suppose that G is an automorphism group of a symmetric design which fixes a block B. Then G permutes the points of B among themselves. Can some nonidentity element of G act as the identity on the points of B, fixing them all? In the language of permutation groups, can G ever act unfaithfully on B? The answer in general is yes. As an example, consider PG(m,2). A block B corresponds to an m-dimensional subspace W of an (m+1)-dimensional vector space V over $\mathbb{F}_2$. It is possible to find a nonidentity automorphism $\sigma \in PGL(m+1,2)$ which fixes every element of the hyperplane W; indeed, there are 2^m-1 such automorphisms. (Under appropriate hypotheses, however, we can show that G must act faithfully on B. See Problem 2.)

We might digress to notice that this automorphism σ of PG(m,2) is interesting because it fixes an exceptionally large number of points--namely $\frac{1}{2}(v-1)$. The following theorem of Feit [38] shows that this is almost the maximum attainable

Theorem 3.5. Suppose that a nonidentity automorphism σ of a nontrivial symmetric (v,k,λ) design fixes f points. Then $f \le \frac{1}{2}v$.

Moreover, if equality holds, then $v=4n$ and σ must be an involution.

Remark. Feit's original proof was quite complicated In its place, I offer a lovely unpublished proof due to H. Wilbrink, which he has kindly allowed me to include.

Proof. After possibly replacing the design by its mplement, we may assume that $k<\frac{1}{2}v$ and $k>2\lambda$.

Suppose that σ has v_1 non-fixed points and v_1 non-xed blocks. Let k_1 be the average number of non-fixed ints on a non-fixed block. If B is a non-fixed block, then e symmetric difference $B\Delta\sigma B$ contains only non-fixed points. nce $v_1 \geq |B\Delta\sigma B| \geq 2n$. Moreover B has at least n non-fixed ints by this argument. Thus $k_1 \geq n$. We now invoke a result e to W. Haemers, which is proven in a set of Supplementary oblems to this chapter.

Lemma 3.6. Let D be a symmetric (v,k,λ) design. ppose that F is a substructure with v_1 points, v_1 blocks d an average of k_1 points on a block. Then

$$n \geq \left(\frac{k_1 v - k v_1}{v - v_1}\right)^2 .$$

oreover, if equality holds, then every block has exactly k_1 ints in the substructure.

Applying the lemma to the substructure of non-fixed ints and non-fixed blocks, and using the inequality $k_1 \geq n$, have

$$\sqrt{n} \geq \frac{k_1 v - k v_1}{v - v_1} \geq \frac{nv - kv_1}{v - v_1} .$$

ter rearranging terms, we have

$$v_1 \geq \left(\frac{n - \sqrt{n}}{k - \sqrt{n}}\right) v .$$

anging this inequality and the inequality $v_1 \geq 2n$ into statents about $f=v-v_1$, we obtain,

$$f \leq (1- \frac{2n}{v})v \quad \text{and} \quad f \leq \left(\frac{\lambda}{k - \sqrt{n}}\right) v$$

assert that the coefficient of v in at least one of the equalities is less than or equal to $\frac{1}{2}$. For, suppose that

$$\frac{\lambda}{k - \sqrt{n}} \geq \tfrac{1}{2}.$$

Then $\sqrt{n} \geq (k-2\lambda) \geq 0$ and thus $(k-\lambda) \geq (k-2\lambda)^2$. Expanding and rearranging we obtain

$$4n \geq \frac{k(k-1)}{\lambda} + 1 = v.$$

Thus,

$$(1 - \frac{2n}{v}) \leq \tfrac{1}{2}.$$

Thus shows that $f \leq \frac{1}{2}v$.

In fact, if equality holds then every "greater-than-or-equal-to" sign becomes "equal-to" in the argument above. Thus $v=4n$. Also $k_1=n$. And, by the lemma, every non-fixed block contains exactly n non-fixed points and λ fixed points. Suppose that σ is not an involution. Then we may choose a block B such that B, σB and $\sigma^2 B$ are distinct. But this accounts for $3n=3/4v$ non-fixed points, which is impossible. Hence σ is an involution.□

For an example meeting the bound, see Problem 13. Our proof actually demonstrates a stronger result, giving an estimate of the number of fixed points of an automorphism.

Corollary 3.7. *Suppose that a nonidentity automorphism σ of a nontrivial symmetric (v,k,λ) design fixes f points. Then*

$$f \leq v-2n \qquad \text{and} \qquad f \leq \left(\frac{\lambda}{k - \sqrt{n}}\right)v.$$

Moreover, if equality holds in either inequality, σ must be an involution and every non-fixed block contains exactly λ fixed points.

Proof. The proof is essentially above. We should simply remark that switching to the complementary design, if $k>\frac{1}{2}v$, only improves the latter bound.□

One important strategy in studying possible symetric designs with particular parameters is to examine
ssible automorphisms and their fixed points and blocks.
r example, suppose that there exists a symmetric (111,11,1)
esign--that is, a projective plane of order 10--and that it
lmits an automorphism σ of prime order p. If σ fixes no
ints or blocks then p divides 111; that is, p=3 or 37.
ippose on the other hand that σ fixes at least one block B.
σ acts faithfully on B then σ induces a nontrivial per-
itation of the 11 points of B; thus $p \leq 11$. If σ acts
ifaithfully on B then it can fix at most one point off B.
heck that otherwise σ would necessarily fix every point.)
ierefore there could be either 99 or 100 non-fixed points.
ince p is prime, again we have $p \leq 11$ in this case. The only
ssible values for p then are p=2,3,5,7,11 or 37. A number
authors have attacked these cases over the last quarter
entury and recently all have been shown to be impossible.
a projective plane of order 10 exists it therefore must
ive only the identity automorphism. (Later on we shall see
w to exclude p=7,11 and 37 and to narrow down the possible
ructure of an automorphism with p=5.) Recently, Janko
id Van Trung have begun work on a putative plane of
der 12 and have shown that an automorphism of prime
der in such a plane must have order 2 or 3. See [67,68,69].

In a similar spirit but with a more positive
esult, Hall [51] has used the information about possible
itomorphisms of a symmetric (41,16,6) design to produce the
irst example of such a design.

3.2 DOUBLY-TRANSITIVE SYMMETRIC DESIGNS

Only a handful of symmetric designs are known which
lmit 2-transitive automorphism groups. These "2-transitive
ymmetric designs" are organized into four classes: two
ifinite classes and two sporadic examples. The designs
ijoy interesting properties, many of which characterize
ie designs completely.

• *Class I*. The symmetric designs PG(m,q) and their complements admit 2-transitive automorphism groups. Many characterizations of these designs are known of which we mention three.

Theorem 3.8. (Ostrom-Wagner [106]) *A projective plane with a 2-transitive automorphism group is isomorphic to* PG(2,q), *for some prime power* q.

Theorem 3.9. (Kantor [74]) *A 2-transitive symmetric design having the same parameters as* PG(m,q) *must be isomorphic to* PG(m,q).

Theorem 3.10. (Ito [66]) *A 2-transitive symmetric design with a nonidentity automorphism fixing every point on some block must be isomorphic to some* PG(m,q).

• *Class II*. The unique symmetric (11,5,2) design H(11) and its complement admit a doubly transitive automorphism group of order 660 (see Problem 4 of Chapter 1). The group is isomorphic to PSL(2,11). The group PSL(2,q) is usually seen in its 2-transitive representation on the q+1 poi of PG(1,q). In the case of H(11), we have one of the "exceptional" representations--of which there are six--of a PSL(2,q) acting transitively in some other way on a "small" number of points. (See [85].)) We have the following characterizations

Theorem 3.11. (Kantor [74]) *A 2-transitive symmetric* (v,k,λ) *design for which* n *is prime is isomorphic to either* PG(2,n), H(11) *or their complements*.

Theorem 3.12. (Kantor [74]) *A 2-transitive symmetric* (v,k,λ) *design for which* k *is prime is isomorphic to some* PG(m,q) *or to* H(11).

• *Class III*. The third class of symmetric designs have parameters $(2^{2m}, 2^{2m-1} \pm 2^{m-1}, 2^{2m-2} \pm 2^{m-1})$ and are best constructed by using quadratic forms over F_2. We mentioned in Appendix B that whereas, over fields of characteristic $\neq 2$, a quadratic form can be recovered from its scalar product (by $Q(x) = \frac{1}{2}B_Q(x,x)$), in characteristic 2 this is no longer possible. In general, many quadratic forms induce the same scalar product. For the field F_2, the relationship between quadratic forms and scalar products is particularly simple.

A quadratic form over an F_2-vector space V is simply map Q: $V \longrightarrow F_2$ satisfying:

(1) $Q(\underline{0}) = 0$, where $\underline{0}$ is the zero vector.

(2) the map B(Q), given by $B(Q)(x,y) = Q(x+y) + Q(x) + Q(y)$ is an (alternating) bilinear form.

Let $\mathcal{Q}$ be the set of quadratic forms on V and let be the set of alternating bilinear forms. The sum of two ıadratic forms is again a quadratic form and $\mathcal{Q}$ has the tructure of an F_2-vector space under addition. (By considerıg polynomials $\Sigma a_{ij}x_ix_j$, show that $\mathcal{Q}$ has dimension $\frac{1}{2}s(s+1)$, ıere $s = \dim V$.) For $B \in \mathcal{B}$, let $\mathcal{Q}_B = \{Q \in \mathcal{Q} \mid B(Q)=B\}$. If B_0 s the zero bilinear form, then $\mathcal{Q}_{B_0}$ is the set of all linear ınctionals on V. So, $\mathcal{Q}_{B_0}$ is a subspace of $\mathcal{Q}$ of dimension s. vo quadratic forms Q_1 and Q_2 lie in the same set $\mathcal{Q}_B$ if and ıly if $Q_1 + Q_2 \in \mathcal{Q}_{B_0}$. Thus, the sets $\mathcal{Q}_B$ are precisely the osets of $\mathcal{Q}_{B_0}$ in $\mathcal{Q}$. Hence there are 2^s quadratic forms ssociated with any given bilinear form B. Two such forms iffer by a linear functional.

The construction of the symmetric design proceeds s follows. Let V be an F_2-vector space of even dimension =2m. Let Q be a nonsingular quadratic form (i.e., B(Q) is onsingular). For the sake of concreteness the reader might ake

$$Q((x_1,\ldots,x_{2m})) = x_1x_2 + \ldots + x_{2m-1}x_{2m}.$$

ıe points of the design are the elements of V. The blocks f the design are indexed by the points. For $u \in V$, the block $_u$ consists of the zeroes of the function $F_u(x) = Q(x+u)$. ıus, for example, $X_{\underline{0}}$ consists of the isotropic points of Q. ore generally,

$$F_u(x) = Q(u) + [Q(x) + B(x,u)].$$

he function $x \longmapsto Q(x) + B(x,u)$ is a quadratic form in $\mathcal{Q}_B$. ıus the points of X_u are either the isotropic vectors of a particular form in Q_B (in the event that $Q(u) = 0$) or the nonisotropic vectors (if $Q(u) = 1$).

Why does this define a symmetric design?

(1) Certainly, there are an equal number of points and blocks; $v=b=2^m$.

(2) The number of points on a block is constant; call it k. Also, since $F_u(x) = F_x(u)$, the number of blocks containing a point x is

$$|\{u|F_u(x) = 0\}| = |\{u|F_x(u) = 0\}| = k.$$

(3) Suppose that two distinct blocks, X_u and X_w, meet in λ points. The number of points contained in exactly one of them is $2(k-\lambda)$. This number is precisely the number of points at which F_u and F_w disagree. As we noted above, $F_u(x) = Q_1(x) + \varepsilon_1$ and $F_w(x) = Q_2(x) + \varepsilon_2$, where Q_1 and Q_2 are quadratic forms in Q_B and where $\varepsilon_1, \varepsilon_2 \in \{0,1\}$. Thus

$$|\{x|F_u(x) \neq F_w(x)\}| = |\{x|Q_1(x)+Q_2(x) \neq \varepsilon_1+\varepsilon_2\}|.$$

Since $Q_1 + Q_2$ is a linear functional (nonzero since $u \neq w$), there are $2^{2m-1} = \frac{1}{2}v$ solutions to the equation $(Q_1 + Q_2)(x) = \Sigma_1 + \Sigma_2$. Thus $v=4(k-\lambda)$ and λ is independent of the choice of blocks. We have now shown that our incidence structure is a symmetric design.

In fact, the symmetric design has $v=4n$ and so, by our observations concerning H-designs in Chapter 1, the parameters must be

$$(v,k,\lambda) = (2^{2m},\ 2^{2m-1} \pm 2^{m-1},\ 2^{2m-2} \pm 2^{m-1}).$$

(There are two other ways of determining the parameters. We could explicitly count the number of isotropic vectors of a quadratic form, with the ± sign depending on which of the two inequivalent types of quadratic forms are chosen. Or, we could simply invoke Problem 3 of Chapter 2.)

We denote the symmetric design with parameters $(2^{2m},\ 2^{2m-1} + 2^{m-1},\ 2^{2m-2} + 2^{m-1})$ by $S^+(2m)$ and its complement by $S^-(2m)$.

Let G be the automorphism group of $S^+(2m)$. For each $u \in V$, the translation map f_u sending $x \longmapsto x + u$ defines an automorphism of $S^+(2m)$. Together the group Σ of all translations is a subgroup of G acting regularly on points (and blocks). To show that G is 2-transitive we must show that $G_{\underline{0}}$, the stabilizer of the zero vector, is transitive on $V-\{\underline{0}\}$. To see this, choose any two points y, $z \in V-\{\underline{0}\}$ and some block X of $S^+(2m)$ containing them. By construction, the points of X are either the isotropic or non-isotropic points of some $Q \in \mathcal{Q}_B$. By the Witt Extension Theorem (Appendix B), the orthogonal group of Q contains some element η carrying y to z. Now, η induces an automorphism of $S^+(2m)$ contained in $G_{\underline{0}}$. Hence $G_{\underline{0}}$ is transitive on nonzero vectors and G is indeed 2-transitive.

In fact, $G_{\underline{0}}$ consists precisely of the isomorphisms $:V \longrightarrow V$ which preserve the alternating bilinear form B. (Prove this.) This is the so-called *symplectic group*, Sp(2m,2).

We mention three characterizations of the $S^{\pm}(2m)$, due to Kantor [74], proving only the last. (The reader should verify that the $S^{\pm}(2m)$ actually possesses the properties asserted. In this connection, see Problem 13.)

Theorem 3.14. *Let* D *be a symmetric design admitting an automorphism group* G *which is* 2-*transitive and which contains a regular normal subgroup. Suppose that for every block* B, *the group* G_B *acts* 2-*transitively on* B *and on the complement of* B. *Then* D *is isomorphic to some* $S^+(2m)$ *or* $S^-(2m)$.

Theorem 3.15. *Let* D *be a symmetric design. The following are equivalent*:

(1) *For any distinct blocks* B *and* C, *there is a nontrivial automorphism of* D *fixing all points not in the symmetric difference* $B \Delta C$.

(2) D is isomorphic to either $S^+(2m)$, $S^-(2m)$, PG(m,2) or complement of PG(m,2), for some integer m.

Theorem 3.16. Suppose that D is a symmetric design admitting a 2-transitive automorphism group, containing a nontrivial element fixing at least $\frac{1}{2}v$ points. Then D is isomorphic to some $S^+(2m)$ or $S^-(2m)$.

Proof. By Theorem 3.5, the element γ fixing $\frac{1}{2}v$ points is an involution and we must have v=4n. If g moves the block B, then the complement of $B\Delta gB$ is the set of fixed points of g. Since D has a 2-transitive automorphism group, Theorum 3.15 applies. □

• Class IV. The most recently discovered class consists of a remarkable symmetric (176,50,14) design D_{176} and its complement. We shall construct it by using the 5-(24,8,1) design D_{24} which we discussed in Chapter 2. Let x and y be distinct points of D_{24}. We say that a block of D_{24} is of Type I if it contains x but not y and Type II if it contains y but not x. There are exactly 176 blocks of each type. (For, check that λ_{12}=176 in the intersection triangle in Problem 18 of Chapter 1.) Recall that distinct blocks of D_{24} meet in 0,2 or 4 points. Given a block B of Type I,

(1) exactly 15 blocks of Type II meet B in 0 points (since μ_0=15 in part (v) of Problem 18 of Chapter 1).

(2) exactly 35 blocks of Type II meet B in 4 points (since there are $\binom{7}{4}$=35 ways to choose the 4 points of B-{x} and there is exactly one block containing these 4 points and the point y).

(3) the remaining 126 blocks of Type II meet B in exactly 2 points.

We define an incidence structure D_{176} as follows: the blocks are the Type I blocks of D_{24}, the points are the Type II blocks of D_{24} and a point and block are incident in D_{176} if the corresponding blocks of D_{24} meet in 0 or 4 points. In D_{176}, every point is incident with 50(=15+35) blocks and every block is incident with 50 points. Hence, D_{176} is a 1-design. To show that it is a 2-design (and hence a symmetric design) we must demonstrate that the number of points ncident with a pair of distinct blocks is a constant, ndependent of the particular pair of blocks.

Notice that the Mathieu group M_{22} acts naturally as n automorphism group of D_{176} (in that it acts as an auto- orphism group of D_{24} stabilizing each of the points x and y). n fact, M_{22} acts as a rank 3 permutation group on the set S f blocks of Type I in D_{24}. That is, M_{22} has three orbits on x S--namely, ordered pairs of blocks meeting in 2,4, or 8 oints. (See Problem 17.) Thus, there are at most two kinds" of pairs of distinct blocks of Type I in D_{24}. For =2,4, let λ_i be the number of points incident in D_{176} with wo distinct blocks of Type I which share i points. We must how that $\lambda_2=\lambda_4=14$.

The average number $\overline{\lambda}$ of points incident with a pair f distinct blocks in D_{176} is easily found to be 14. More- ver by exploiting the geometry of D_{24} developed in Chapter 2, e can show that $\lambda_2=14$. (See Problem 18.) Since $\overline{\lambda}=14$ and $_2=14$, we must also have $\lambda_4=14$. Hence D_{176} is indeed a sym- etric (176,50,14) design.

The happy coincidence that $\lambda_2 = \lambda_4$ makes one wonder hether, from the point of view of D_{176}, there is only one kind" of pair of distinct blocks. That is, it suggests hat D_{176} might have a rank 2, or 2-transitive, automorphism roup. In fact, this is the case. The full automorphism roup of D_{176} is a group 100 times larger than M_{22}. This roup is the Higman-Sims group, which is a simple group of rder 44,352,000. Unfortunately, we do not have the space to igress here to discuss this interesting group.

It would be of great interest to know all 2-transi- ive symmetric designs. It appears that this problem has ecently been essentially solved.

By the following result, the determination of all -transitive symmetric designs will be complete once we know enough about) all 2-transitive permutation groups.

<u>Theorem 3.17</u>. <u>Let</u> D <u>be a nontrivial symmetric</u> v,k,λ) <u>design with a 2-transitive automorphism group</u> G. <u>onsidering</u> G <u>as a permutation group on points, the subgroup</u>

G_B stabilizing a block B (setwise) is an intransitive subgroup of G having index v and having an orbit of size k with $1<k<v-1$.

Conversely, let G be any permutation group acting 2-transitively on a set of v points. Suppose that G has a subgroup H of index v which acts intransitively, having some orbit B of size k, with $1<k<v-1$. The images of B under G define the blocks of a symmetric design admitting G as an automorphism group.

Proof. The first half is clear. About the second half, we note only that the 2-transitivity of G guarantees that the number of blocks containing a pair of distinct points is constant.□

The determination of all 2-transitive permutation groups is a by-product of the recent classification of finite simple groups. For, the minimal normal subgroup of a 2-transitive group is either a simple group acting transitively or an elementary abelian group acting regularly. (See [141].) The classification makes it possible to determine the first case; work by Hering and others has investigated the second. While not all of the proofs have yet appeared in print, it seems safe to say that all 2-transitive groups and all 2-transitive symmetric designs are known.

Still, solving the problem of determining all 2-transitive symmetric (v,k,λ) designs by invoking the massive work which undergirds the classification of finite simple groups seems unsatisfactory. After all, for the case $\lambda=1$ a much more elementary proof is known (see Theorem 3.8). A simpler, more combinatorial proof will continue to be desirable.

§3.3 AUTOMORPHISMS OF PRIME ORDER

After having examined the "largest" possible automorphism groups of symmetric designs in the last section, we turn now to prove a theorem about the "smallest," namely those of prime order. Our main tools will be the self-dual codes introduced in Chapter 2 and the representation theory of groups. (The reader should at this point become familiar th the contents of Appendix D.) Throughout the discussion, x a symmetric (v,k,λ) design D and an automorphism group G.

We defined the module M of the symmetric design D be the integral span of the rows of the extended incidence trix

$$B = \left(\begin{array}{c|c} A & \begin{matrix}1\\ \vdots \\ 1\end{matrix} \\ \hline \lambda \dots \lambda & k \end{array}\right).$$

can regard M as a submodule of $W = \mathbb{Z}^{v+1}$, the collection all $(v+1)$-tuples of integers. Suppose that the i-th ordinate of W corresponds to the point p_i of D (for $i=1,$ $\ldots,v)$. We can define a natural action of G on W as follows: $w=(w_1,\ldots,w_{v+1})$, we let $gw=(x_1,\ldots,x_{v+1})$ where

$$x_i = \begin{cases} w_{\eta(i)} & \text{for } i=1,\ldots,v; \\ w_{v+1} & \text{for } i=v+1. \end{cases}$$

ere $p_i \longmapsto p_{\eta(i)}$ is the point permutation defined by g. In ort, g permutes the coordinates of W as it permutes the rresponding points. In this way, W is a representation dule for G or, equivalently, W is a (left) $\mathbb{Z}G$-module. tice that if w happens to be a row of B then gw is also a w of B. Hence, the module M is closed under the action of That is, M is a sub-$\mathbb{Z}G$-module of W.

In a similar fashion, the codes C_α which we defined Chapter 2 can be viewed as representation modules for G. t p be a prime. View $V=F_p^{v+1}$ as the reduction (mod p) of The $\mathbb{Z}$-module

$$p^{-\alpha}M \cap W$$

certainly fixed under the action of G. So is its image under reduction (mod p). That is, C_α is a sub-F_pG-module f the F_pG-module V.

Finally, not only does G preserve all the relevant codes and modules, it also preserves scalar products of the sort we have been using. If ψ is a scalar product of the form

$$\psi(x,y) = x_1y_1+\ldots+x_vy_v - ax_{v+1}y_{v+1}$$

for $x = (x_1,\ldots,x_{v+1})$ and $y = (y_1,\ldots,y_{v+1})$ then

$$\psi(x,y) = \psi(gx,gy)$$

for all $g \in G$ and x,y in whatever code or module is under discussion.

Our strategy in this section will be to exploit the self-dual codes associated with certain symmetric designs in just the same manner as in Chapter 2. There we used the fact that the very existence of a self-dual F_p-code entailed restrictions on the vector space V and its scalar product ψ (namely, that $((-1)^{\frac{1}{2}\dim V}\det\psi)$ is a square (mod p)). Here we use the fact that the existence of such a self-dual F_p-code which is fixed by a group G acting on V places restrictions on the structure of V as a representation module for G. The restrictions are on the composition factors of V as an F_pG-module.

Proposition 3.18. *Suppose that*

(1) G *is a finite group*,

(2) F *is a field*,

(3) V *is an* FG-*module, finite-dimensional as a vector space over* F,

(4) W *is a sub*-FG-*module of* V,

(5) ψ *is a nondegenerate scalar product on* V *such that* $\psi(x,y) = \psi(gx,gy)$ *for all* $g \in G$ *and* $x,y \in V$, *and*

(6) $W = W^{\psi}$

For any irreducible FG-*module* T *which occurs as a composition factor of* V,

(a) *if* $T \neq T^*$ *then* T *and* T^* *occur with equal multiplicity as composition factors of* V;

(b) *if* $T \simeq T^*$ *then* T *occurs with even multiplicity as a composition factor of* V.

Remark. By T^* we denote $\mathrm{Hom}_F(T,F)$, the contra-
edient representation, which is defined and endowed with a
ft FG-module structure in Appendix D.

Proof. We want to fill out the chain

$$0 \subseteq W \subseteq V$$

obtain a composition series for V as an FG-module. Suppose
at

$$0 = W_0 \subseteq W_1 \subseteq \ . \ . \ . \subseteq W_{t-1} \subseteq W_t = W$$

a chain of FG-modules which is a composition series for W.
at is, W_i/W_{i-1} is an irreducible FG-module for i=1,...,t.
en

$$0 = W_0 \subseteq \cdots \subseteq W_{t-1} \subseteq W{=}W^{\psi} \subseteq W_{t-1}{}^{\psi} \subseteq \cdots \subseteq W_0{}^{\psi} = V.$$

a composition series for V. (To prove this involves two
eps. First show that $W_i{}^{\psi}$ is actually a module closed under
e action of G, by using assumption (6). Then note that no
-module X can be "squeezed" in on the right half of the
ain, for otherwise X^{ψ} could be inserted on the left
lf.) Next we prove a lemma which, while essentially
ementary, is an exercise in formal homological algebra.

Lemma 3.19. *The isomorphism of* FG-*modules*

$$(W_i/W_{i-1})^* \simeq (W_{i-1}{}^{\psi}/W_i{}^{\psi})$$

lds for i=1,...,t.

Proof of Lemma. Define a map Φ: $W_{i-1}{}^{\psi} \longrightarrow \mathrm{Hom}_F(W_i,F)$
follows. For $x \in W_{i-1}{}^{\psi}$, set Φ to be the linear functional
ch that

$$\Phi(x)(y) = \psi(x,y)$$

for $y \in W_i$. This is a homomorphism of FG-modules since $\Phi(gx)(y) = \psi(gx,y) = \psi(x,g^{-1}y) = \phi(x)(g^{-1}y) = (g\Phi(x))(y)$ and thus $\phi(gx) = g\phi(x)$. (Recall how we made $\mathrm{Hom}_F(W_i,F)$ a left FG-module.)

Now, define a map $\phi':W_{i-1}{}^{\psi} \longrightarrow \mathrm{Hom}_F(W_i/W_{i-1},F)$ as follows. If $\bar{y} = y + W_{i-1}$ is an element of W_i/W_{i-1} set

$$\phi'(x)(\bar{y}) = \phi(x)(y).$$

(Check that Φ' is well-defined.) Next notice that the kernel of Φ' is precisely $W_i{}^{\psi}$. Hence $W_{i-1}{}^{\psi}/W_i{}^{\psi}$ is isomorphic to a submodule of $\mathrm{Hom}_F(W_i/W_{i-1},F)$. We need only show that the submodule must in fact be all of $\mathrm{Hom}_F(W_i/W_{i-1},F)$.

To see this it is sufficient to notice that as F-vector spaces $\dim_F(W_{i-1}{}^{\psi}/W_i{}^{\psi}) = \dim_F(\mathrm{Hom}_F(W_i/W_{i-1},F))$. Hence,

$$(W_{i-1}{}^{\psi}/W_i{}^{\psi}) \simeq \mathrm{Hom}_F(W_i/W_{i-1},F) = (W_i/W_{i-1})^*,$$

which completes the proof of the lemma.□

Applying the lemma, we see that whenever a composition factor T occurs in the first half of the composition series for V, then T* occurs in the second half, and _vice versa_. This pairing establishes the proposition.□

Remark. We can weaken the hypotheses of Proposition 3.18 at the expense of greatly weakening the conclusion. Specifically retain hypotheses (1) - (5), but replace (6) by $W \subseteq W^{\psi}$. The above proof is sufficient to show that if T is a composition factor of W occurring with multiplicity r and if $T \simeq T^*$, then T occurs as a composition factor of V with multiplicity at least 2r. We shall use this observation in Chapter 4.

The next theorem shows how Proposition 3.18 yields a necessary condition for a symmetric design to possess an automorphism group of odd prime order q.

Theorem 3.20. _Suppose that a symmetric_ (v,k,λ) _design_ D _admits an automorphism group_ G _of odd prime order_ q.

t f be the number of fixed points of G and let w + f be the tal number of orbits of G on points. Then

(1) Either n is a square or w + f is odd.

(2) Suppose that for some prime p dividing n* we have $p^j \equiv -1 \pmod q$ for some integer j. Then w must be even and f must be odd.

Proof. Every point orbit of G has size 1 or size There are f orbits of size 1 and w orbits of size q. nce v = f + wq. If w + f is even then v = w + f + (q-1)w also even. By Schutzenberger's Theorem, n is a square. is proves (1).

Suppose that p is a prime such that $p^j \equiv -1 \pmod q$ r some integer j. (In particular, $p \neq q$.) Consider the ctor space $V = F_p^{v+1}$ to be a representation module for G, the manner described above. We can decompose V as a rect sum of F_pG-submodules corresponding to the coordinates each orbit of G.

$$V \simeq \underbrace{F_p \oplus \dots \oplus F_p}_{f+1 \text{ times}} \oplus \underbrace{F_pG \oplus \dots \oplus F_pG}_{w \text{ times}}$$

ere the f+1 trivial representations correspond to the f xed points of D and the final coordinate of V, which is also xed by G. Now, each composition factor of F_pG occurs with ltiplicity one (since G is abelian and p does not divide |). So, if T is any nontrivial composition factor of V, then occurs as a composition factor of V with multiplicity w. reover, by Appendix D, if $p^j \equiv -1 \pmod q$ then every composi- on factor of F_pG is self-contragredient.

Since p divides n*, we can find an F_p-code C sociated with D which is self-dual with respect to an propriate scalar product ψ, by §2.3. Proposition 3.18 now plies to V. Hence, every self-contragredient composition ctor of V occurs with even multiplicity. Thus w is even. nce w + f is odd, f is odd.□

As an application, we show that a projective plane of order 10 cannot have automorphisms of order 37 or 7. We have n=10 and both 2 and 5 divide n*.

(1) Since $2^{18} \equiv -1 \pmod{37}$, any automorphism of order 37 fixes an odd number of points, by Theorem 3.20. However, by the discussion in §3.1, such an automorphism must fix no points. Contradiction.

(2) Consider an automorphism of order 7. Since v=111, the number of fixed points f satisfies $f \equiv 111 \equiv 6 \pmod 7$. Since $5^3 \equiv -1 \pmod 7$, then f must be odd by Theorem 3.20. Hence, $f \equiv 13 \pmod{14}$. However, an automorphism of a projective plane fixes at most 12 points by Problems 4 and 5. Contradiction.

See Problems 7,21 and 22 concerning automorphisms of order 2,3,5 or 11.

§3.4 COUNTING ORBITS

Leaving aside representation theory temporarily, we shall explore below how the group actions introduced in the previous section can be manipulated in a rather more combinatorial manner to prove results about the orbit structure of an arbitrary automorphism group. Fix the following information: let G be an automorphism group of a symmetric (v,k,λ) design D and suppose that G has r orbits $\mathcal{P}_1,\ldots,\mathcal{P}_r$ on points and r orbits $\mathcal{B}_1,\ldots,\mathcal{B}_r$ on blocks. Further, write $s_i = |\mathcal{P}_i|$ and $t_i = |\mathcal{B}_i|$ for $i=1,\ldots,r$.

The group G acts on the coordinates of the $\mathbb{Z}$-module $W=\mathbb{Z}^{v+1}$, having r+1 orbits. The first r orbits correspond to the $\mathcal{P}_i$ (for $i=1,\ldots,r$) and the last orbit, which we label $\mathcal{P}_{r+1}$, corresponds to the final coordinate. Consider the vector u_i which has a 1 in every coordinate of $\mathcal{P}_i$ and a 0 elsewhere. Clearly u_i is left fixed by every element of G. In fact $u_1,\ldots,u_{r+1}$ is a basis for the submodule W_G consisting of all elements of W left fixed by G;

$$W_G = \{w \in W \mid gw = w \text{ for all } g \in G\}.$$

t ψ be the scalar product on W defined by $\psi(x,y) = y_1+\ldots+x_v y_v - \lambda x_{v+1}y_{v+1}$ for $x = (x_1,\ldots,x_{v+1})$ and $= (y_1,\ldots,y_{v+1})$. If we restrict ψ to the submodule W_G en $\{u_1,\ldots,u_{r+1}\}$ is an orthogonal basis. Indeed,

$$\psi(u_i,u_j) = \begin{cases} 0 & \text{if } i\neq j, \\ s_i & \text{if } i=j\leq r, \\ -\lambda & \text{if } i=j=r+1. \end{cases}$$

Now let M be the $\mathbb{Z}$-module of D and let M_G be the bmodule of M consisting of all elements left fixed by every ement of G. That is, $M_G = M \cap W_G$. If we denote the rows of e extended incidence matrix B by $e_1,\ldots,e_{v+1}$, respectively, en it is not hard to see that M_G is generated by the ements $f_1,\ldots,f_{r+1}$, where f_i is the sum of all e_j correonding to blocks in the orbit $\mathcal{B}_i$. (Here, as with points, t the final row of B lie in a singleton orbit called $_{r+1}$.) Since we know the value of $\psi(e_i,e_j)$ for $1\leq i,j\leq v+1$, can compute that

$$\psi(f_i,f_j) = \begin{cases} 0 & \text{if } i\neq j \\ t_i n & \text{if } i=j\leq r \\ -\lambda n & \text{if } i=j=r+1. \end{cases} \qquad (*)$$

Each f_i can be written as an integral linear comination of the u_i; say, $f_i = \sum b_{ij}u_j$. The matrix $B_G = (b_{ij})$ then the matrix giving the coordinates of $\{f_1,\ldots,f_{r+1}\}$ ith respect to the basis $\{u_1,\ldots,u_{r+1}\}$. Since the matrix f ψ on W_G is the diagonal matrix

$$\psi_G = \begin{pmatrix} s_1 & & & 0 \\ & \ddots & & \\ & & s_r & \\ 0 & & & -\lambda \end{pmatrix}$$

ith respect to $\{u_1,\ldots,u_{r+1}\}$, we can express (*) by the atrix equation,

$$B_G \begin{pmatrix} s_1 & & & & 0 \\ & \cdot & & & \\ & & \cdot & & \\ & & & \cdot & \\ & & & s_r & \\ 0 & & & & -\lambda \end{pmatrix} B_G^T = \begin{pmatrix} t_1 n & & & & 0 \\ & \cdot & & & \\ & & \cdot & & \\ & & & \cdot & \\ & & & t_r n & \\ 0 & & & & -\lambda n \end{pmatrix} . \qquad (**)$$

This equation provides an analogue to the equation $B\psi B^T = n$, which has been our mainstay until now. We call B_G the <u>contraction</u> of B with respect to G. Almost everything we did to B, we can do to B_G. For example, the following analogue to Schutzenberger's Theorem follows at once.

<u>Theorem 3.21</u>. <u>If an automorphism group of a symmetric</u> (v,k,λ) <u>design has orbits of size</u> $s_1,\ldots,s_r$ <u>on points</u> $t_1,\ldots,t_r$ <u>on blocks then</u>

$$\frac{n^{r+1}\, t_1 \ldots t_r}{s_1 \ldots s_r}$$

<u>must be the square of an integer</u>.

An automorphism group is said to be <u>semi-standard</u> if, after possibly renumbering orbits, we have $s_i = t_i$, for $i=1,\ldots,r$. Semi-standard automorphism groups are more common than one might at first imagine. For example, if every point orbit of a group G has either 1 or $|G|$ points then G must be semi-standard. (See Problem 10.) Such a group is called <u>standard</u>; automorphism groups of prime order are of course standard and hence semi-standard. Cyclic automorphism groups need not be standard, but it is not hard to show that they must be semi-standard (Problem 20). For semi-standard automorphism groups we can simplify Theorem 3.21.

<u>Theorem 3.22</u>. <u>If</u> G <u>is a symmetric automorphism group of a symmetric</u> (v,k,λ) <u>design then either</u> n <u>is a square or</u> G <u>has an odd number of orbits on the points of the design</u>.

When G consists of the identity automorphism alone, Theorem 3.22 reduces precisely to Schutzenberger's Theorem. (For an application of Theorem 3.22 see Problem 23.)

In exactly the spirit of §2.3, we can also use equation (**) to produce a chain of F_p-codes--including a self-dual code, under appropriate hypotheses. Provided that p does not divide $s_1 \ldots s_r t_1 \ldots t_r$ the method of §2.3 carries over precisely. We omit the proof.

Theorem 3.23. Suppose that G is an automorphism group of a symmetric (v,k,λ) design D. Let p be a prime which divides n but which does not divide the size of any point or block orbit of G. Then,

(1) if $p|n^*$ and $p\nmid\lambda^*$, there exists an F_p-code C of length r+1 which is self-dual with respect to the scalar product $\psi(x,y) = s_1x_1y_1+\ldots+s_rx_ry_r - \lambda^*x_{r+1}y_{r+1}$. In particular,

$$(-1)^{\frac{1}{2}(r+1)} \det \psi = (-1)^{\frac{1}{2}(r-1)} s_1 \ldots s_r \lambda^*$$

must be a square (mod p).

(2) if $p|n^*$ and $p|\lambda^*$, there exists an F_p-code C of length r+1 which is self-dual with respect to the scalar product $\psi(x,y) = s_1x_1y_1+\ldots+s_rx_ry_r + (\lambda^*n^*/p^2)x_{r+1}y_{r+1}$. In particular,

$$(-1)^{\frac{1}{2}(r+1)} \det \psi = (-1)^{\frac{1}{2}(r+1)} s_1 \ldots s_r (\lambda^*n^*/p^2)$$

must be a square (mod p).

Remarks. (1) The theorem always applies in the case hat p does not divide the order of G, for then no orbit has ize divisible by p.

(2) Actually, we need not require that $p\nmid s_1 \ldots$ $_r t_1 \ldots t_r$ in order to mimic the proof of Theorem 2.17. It s enough that for some integer α, we have $p^\alpha || s_i$ and $p^\alpha || t_i$ for i=1,...,r). For we may then cancel p^α from both sides f (**) and proceed as above with $s_i' = p^{-\alpha}s_i$ and $t_i' =$ $^{-\alpha}t_i$.

If we apply Theorem 3.23 to the case of a standard automorphism group of even order, we obtain the following result.

Proposition 3.24. Suppose that G is a standard automorphism group of a symmetric (v,k,λ) design D. Let f be the number of fixed points of G and let $w + f$ be the total number of orbits of G on points. If G has even order, then:

(1) Either n is a square or w is even.

(2) If some prime p dividing n^* is congruent to 3 (mod 4) then w is a multiple of 4.

Proof. By Theorem 3.22, the number $w + f$ of orbits is odd. If w is odd then f must be even and hence $v=w|G|+f$ is even. By Schutzenberger's Theorem, n must be a square. This proves the first assertion.

Consider a Sylow 2-subgroup of G. It also is a standard automorphism group and it has $w'=dw$ orbits of size greater than 1, where d is the index of the Sylow 2-subgroup in G. Since d is odd, $w \equiv 0 \pmod 4$ if and only if $w' \equiv 0 \pmod 4$. Hence, there is no harm in assuming that the group G in question is a 2-group.

Let p be a prime dividing n^* and let ℓ be $(-\lambda^*)$ or $(\lambda^* n^*/p^2)$, according as $p \nmid \lambda^*$ or $p | \lambda^*$. By Theorem 3.23, the quantity

$$(-1)^{\frac{1}{2}(w+f+1)}|G|^w \ell$$

must be a square (mod p). Since w is even $(-1)^{\frac{1}{2}(w+f+1)}\ell$ must be a square (mod p).

Now, every 2-group contains a subgroup of index 2. Let H be a subgroup of index 2 in G. Then H acts as a standard automorphism group having $(2w+f)$ orbits on points. Applying Theorem 3.23 to H, we see that $(-1)^{\frac{1}{2}(2w+f+1)}\ell$ must be a square (mod p).

The ratio of these two squares is $(-1)^{\frac{1}{2}w}$, which must be a square (mod p). If $p \equiv 3 \pmod 4$, then -1 is not a square (mod p). Thus, in this case, $w \equiv 0 \pmod 4$. □

Remark. In the same fashion, we could apply Theorem 3.23 to obtain a necessary condition for the existence of a standard automorphism group of odd order. If we do this, say, or groups of prime order, the existence condition turns out o be weaker than Theorem 3.20 (see Problem 19), hence we hall not record the result. The situation for groups of omposite odd order is no more interesting, since it turns ut that the existence condition for such groups is equiva- ent to the union of the existence conditions for all sub- roups of prime order.

Before closing this chapter, we take a moment to tudy more closely the contraction of incidence matrices, hich we shall frequently use in the next two chapters.

The definition of B_G may seem somewhat abstract, ut it can easily be put in purely computational terms. heck that to find B_G: (1) add together the rows of B in the ame orbit (i.e., find the f_i); (2) notice that all columns n the same orbit are identical (since the f_i are fixed ectors); (3) save one copy of a column from each orbit i.e., express the f_i with respect to the u_i). From this escription it should be clear that the entries of B_G are onnegative integers less than or equal to $|G|$. We record wo important equations.

<u>Proposition 3.25</u>. <u>With the notation above</u>.

$$B_G\psi_G B_G = \begin{pmatrix} t_1 n & & & \\ & \ddots & & \\ & & t_r n & \\ & & & -\lambda n \end{pmatrix} \quad \underline{\text{and}} \quad B_G\psi_G J = \begin{pmatrix} t_1 n & & & \\ & \ddots & & \\ & & t_r n & \\ & & & n(k-1) \end{pmatrix} .$$

<u>Proof</u>. The first equation has already been proven. or the second equation, notice first that

$$B\psi J = \begin{pmatrix} n & & 0 \\ & \ddots & \\ & & n \\ 0 & & n(k-1) \end{pmatrix}$$

here ψ is the underlying scalar product, $\psi(x,y)=x_1y_1+\ldots+ x_vy_v-\lambda x_{v+1}y_{v+1}$. Now the scalar product of a row f_i of B_G ith the all-one vector $\overline{1}$ is given by

$$\psi(f_i,\bar{I}) = \sum_{e_j \in B_i} \psi(e_j,\bar{I}) = \begin{cases} |B_i|n & \text{if } i=1,\ldots,r \\ n(k-1) & \text{if } i=r+1. \end{cases}$$

This proves the second equation.□

The matrix B_G has the form

$$B_G = \left(\begin{array}{c|c} A_G & \begin{matrix} t_1 \\ \vdots \\ t_r \end{matrix} \\ \hline \lambda \ldots \lambda & k \end{array}\right)$$

The matrix A_G is called the <u>contraction</u> by G of A. Of course, A_G can be obtained directly merely by "contracting" A. Often it is more convenient to work with A_G rather than B_G.

<u>Proposition 3.26</u>. <u>With the notation above</u>.

$$A_G \begin{pmatrix} s_1 & & 0 \\ & \ddots & \\ 0 & & s_r \end{pmatrix} A_G^T = \begin{pmatrix} t_1n+t_1^2\lambda & t_1t_2\lambda & \cdots & t_1t_r\lambda \\ t_2t_1\lambda & \ddots & & \vdots \\ \vdots & & \ddots & \vdots \\ t_rt_1\lambda & \cdots & \cdots & t_rn+t_r^2\lambda \end{pmatrix}.$$

<u>and</u>

$$A_G \begin{pmatrix} s_1 & & 0 \\ & \ddots & \\ 0 & & s_r \end{pmatrix} J = \begin{pmatrix} t_1k & & 0 \\ & \ddots & \\ 0 & & t_rk \end{pmatrix} J.$$

<u>Proof</u>. This follows from Proposition 3.25.□

These formulae become particularly simple in one situation. Suppose that H is an automorphism group acting <u>semiregularly</u> on points and blocks--that is, every orbit has size $h=|H|$. In this case, the equations of the previous proposition reduce to

$$A_H A_H^T = nI + h\lambda J$$

and $$A_H J = kJ.$$

(As in the proof of Theorem 1.10, we also have $JA_H = kJ$.)

Semiregular automorphism groups arise directly from regular automorphism groups. If G is a regular automorphism group of a symmetric design then every subgroup of G is semiregular. Using the techniques above, we shall concentrate on regular automorphism groups in the next two chapters.

PROBLEMS - CHAPTER 3

1. Let σ be an automorphism of a symmetric (v,k,λ) design which fixes f points and F blocks. Count the number of triples $(p,\sigma(p),B)$ where B is a block containing the points p and $\sigma(p)$ first by summing over points and second by summing over blocks. Obtain a purely combinatorial proof that $f=F$.

2. Suppose that D is a symmetric (v,k,λ) design with an automorphism σ fixing every point of some block B. Show that for any other block C, we have $B \cap C = B \cap \sigma(C) = C \cap \sigma(C)$. Use the fact that $|B \cap C \cap \sigma(C)| \leq v$ to show that $k \geq 2\lambda+1$.

3. Suppose that D_1 is a projective plane of order n and that a subset D_2 of the points and blocks (with the restriction of the incidence relation) has the structure of a projective plane of order m, with $m<n$.

(i) Suppose that some block B of D_1 contains no point of D_2. Show that every block of D_2 meets B in one point and that no point of B is incident with more than one block of D_2. Hence show that $n \geq m^2+m$. Dually, if some point of D_1 is incident with no block of D_2, reach the same conclusion.

(ii) Otherwise, every block of D_1 contains a point of D_2 and every point of D_1 is on some block of D_2. Choose a block B of D_1 which has one point of D_2. By a similar counting argument show that $n=m^2$.

4. Suppose that F is a finite incidence structure such that:

(i) any two points are incident with a unique block, and

(ii) any two blocks are incident with a unique point.

Show that one of the following must hold:

(a) F is a projective plane;

(b) some block is incident with every point;

(c) some point is incident with every block; or

(d) F is the incidence structure:

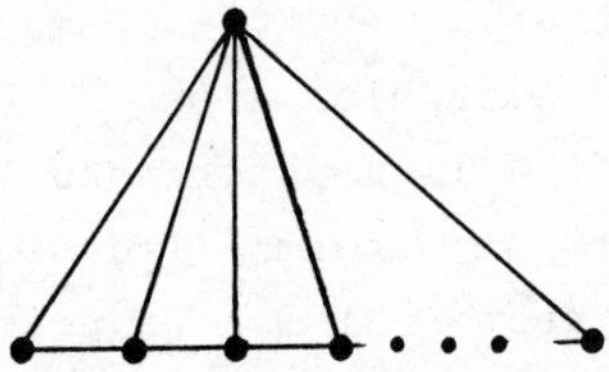

5. A subset of the points and blocks of a projective plane which satisfies the axioms of the previous problem is called a *closed configuration*. Show that the fixed points and blocks of an automorphism of a projective plane form a closed configuration.

6. A closed configuration in a projective plane of order n is called a *Baer configuration* if any point of the plane lies on some block of the configuration and any block of the plane contains some point of the configuration. Show that a Baer configuration C must be one of the following:

(1) For some point p and block B which are incident, C consists of all blocks through p and points on B.

(2) For some point p and block B which are not incident, the blocks of C are B together with all blocks through p and the points of C are p together with all points on B.

(3) A projective subplane of order $\sqrt{n}$.

7. Show that the fixed points and fixed blocks of an involution (automorphism of order 2) in a projective plane of order n form a Baer configuration.

(1) If n is even, show that only types (1) and (3) above can occur.

(2) If n is odd, show that only types (2) and (3) above can occur.

(3) Produce examples of involutions fixing Baer configurations of each of the three types.

(If $n \equiv 2 \pmod 4$), then only type (1) can occur since n is not a square. Hughes [61] proved in fact that type (1) can only occur for n=2. Hence, a projective plane of order n has no involutions if n>2 and $n \equiv 2 \pmod 4$)).

8. Suppose that D is a symmetric (v,k,λ) design with $\lambda \geq 2$.

(1) If G is an automorphism group of D and every prime factor of the order of G is greater than λ, show that the firxed points and blocks of G form a 1-design.

(2) If D has an automorphism σ of prime order p, then either $p|v$ or $p \leq k$. (Hint: if not, then σ fixes every point of a block B. Use (1).)

9. Show that a symmetric design admitting a regular abelian automorphism group is isomorphic to its dual.

10. Suppose that G is an automorphism group of a symmetric design and that every point orbit of G has size 1 or $|G|$. Prove that every block orbit also has size 1 or $|G|$ and that the number of point and block orbits of each size is equal.

11. (1) Show that the full automorphism group of H(q) has rank 2 if q=3,7, or 11 and rank 3 if $q \geq 19$.

(2) Show that the full automorphism group of A(2m,q) has rank 3. (Hint: Use the Witt Extension Theorem.)

12. Let D be a symmetric (v,k,λ) design admitting a 2-transiti automorphism group G.

(1) Show that the stabilizer of a block acts transitively on the points incident with the block and the points not incident with the block.

(2) Show that, if v and k are relatively prime, the subgroup stabilizing a block B and a point $p \in B$ is transitive on the points not incident with B.

13. Find an automorphism of $S^{\pm}(2m)$ fixing $\frac{1}{2}v$ points. (Hint: the symmetric difference of two blocks has $\frac{1}{2}v$ points and corresponds to a translate of a subspace of dimension 2m-1.)

14. Suppose that B is a nondegenerate alternating bilinear form on a vector space of dimension 2m+1 over $\mathbb{F}_2$. Use the argument of §3.2 to produce a symmetric design with an even number of points with n a nonsquare. Thus obtain a (rather underhanded) proof that no such bilinear form exists.

5. Suppose that we use the quadratic form

$$Q(\underline{x}) = x_1x_2+\dots+x_{2m-1}x_{2m}$$

n the construction of $S\pm(2m)$. By viewing Q as a sum of uadratic forms on 2-dimensional vector spaces, show that the esign in Example 10 of Chapter 1, obtained by taking Kronecker roducts of

$$\begin{pmatrix} +1 & -1 & -1 & -1 \\ -1 & +1 & -1 & -1 \\ -1 & -1 & +1 & -1 \\ -1 & -1 & -1 & +1 \end{pmatrix}$$

s isomorphic to $S^{\pm}(2m)$ (with the sign being + or - according s m is even or odd).

6. (1) Show that a regular permutation group on a set of objects can be 2-homogeneous only if $m\leq 4$.

(2) Note that Theorem 3.17 remains true if 2-transitive s replaced by 2-homogeneous. Suppose now that G is a -homogeneous permutation group on a set X of cardinality m. et B be an orbit of G_x, for some $x \in X$. Show that $|B| = 1$ r m-1. (Hint: Problem 12 of Chapter 1.)

(3) Show that a 3-homogeneous permutation group on a set f m objects must be 2-transitive if $m\geq 5$.

he next two problems prove assertions used in the construcion of D_{176}--specifically, that M_{22} acts as a rank 3 perutation group on the blocks of Type I in D_{24} and that the ntersection number $\lambda_4=14$. In view of the construction of $_{24}$ and the 5-transitivity of its automorphism group, if any hree points a,b,c are deleted from D_{24} each block B becomes block, oval, Fano plane or symmetric difference of two locks in the resulting PG(2,4)(according as B meets {a,b,c} n 3,2,1 or 0 points). To solve the next two problems, onvert statements about Type I and Type II blocks into tatements about PG(2,4). The delicate point is which three points of D_{24} to delete to make the resulting statements about PG(2,4) easiest to deduce from the geometry of the plane.

17. Let S be the set of Type I blocks in D_{24}. We wish to show that M_{22} has three orbits on S x S, consisting of pairs of blocks (of Type I) meeting in 2,4, or 8 points.

(1) Prove that M_{22} is transitive on pairs of blocks meeting in 8 points.

(2) Prove that M_{22} is transitive on pairs of blocks meeting in 4 points if and only if M_{22} is 2-transitive on the blocks (of Type I) through any particular 4 points.

(3) Show that the latter condition holds if and only if the automorphism group fixing two points x and y in PG(2,4) acts 2-transitively on the set of blocks through x but not y. Verify this property of PG(2,4).

(4) Prove that M_{22} is transitive on pairs of blocks meeting in 2 points if and only if the full automorphism group of PG(2,4) is transitive on pairs of disjoint ovals. Verify this property of PG(2,4).

18. Let B_1 and B_2 be two blocks of Type I in D_{24} meeting in four points. The parameter λ_4 is the number of blocks of Type II incident in D_{176} with both B_1 and B_2.

(1) Let x and y be two points of PG(2,4) and let ℓ_1 and ℓ_2 be two blocks of PG(2,4) meeting in x and containing y. Show that λ_4 is also the total number of:

(a) blocks incident with y, not incident with x (and necessarily meeting ℓ_1 and ℓ_2 in 1 point each).

(b) ovals incident with y, not incident with x and meeting ℓ_1 and ℓ_2 in 2 points each.

(c) Fano planes incident with y, not incident with x and meeting ℓ_1 and ℓ_2 in 3 points each.

(d) symmetric difference of two blocks, incident with y, not incident with x and meeting ℓ_1 and ℓ_2 in 0 or 4 points each.

(2) Use the geometry of PG(2,4) to show that the
ımber of objects of types (a),(b),(c),(d) is 4,6,0,4,
espectively. Hence λ_4=14.

). Consider an automorphism group of a symmetric design
ıving odd prime order q. By applying Theorem 3.23 obtain
modified version of Theorem 3.20 in which the condition
$^j \equiv -1$ (mod q) for some integer j" is replaced by "p is not
square (mod q)." Show that the latter condition implies
ıe former but not conversely and thus that the modified
ersion is weaker.

). Show that a cyclic automorphism group is semi-standard.

l. Let σ be an automorphism of a projective plane of order
). Show that

(1) if σ has order 3, then σ fixes 3 or 9 points;

(2) if σ has order 5, then σ fixes 1 or 11 points;

(3) if σ has order 11, then σ fixes 1 point.

2. Suppose that σ is an automorphism of order 11 of a pro-
ective plane of order 10. Suppose that σ fixes the point
and block b. Observe that p∉b. Label the row orbits of
σ> on the extended incidence matrix of the design so that
$_{10}$ is the orbit containing b alone, B_{11} is the orbit
ontaining the blocks on p and B_{12} is the orbit corresponding
o the final row. Label the column orbits similarly with
$_1, \ldots, P_{12}$. Observe that the contraction $B_{<\sigma>}$ has the form

$$B_{<\sigma>} = \left(\begin{array}{cccc|ccc} & & & & 0 & 1 & 11 \\ & & & & \cdot & \cdot & \cdot \\ & & X & & \cdot & \cdot & \cdot \\ & & & & \cdot & \cdot & \cdot \\ & & & & 0 & 1 & 11 \\ \hline 0 & \cdot & \cdot & \cdot\, 0 & 0 & 1 & 1 \\ 1 & \cdot & \cdot & \cdot\, 1 & 11 & 1 & 11 \\ 11 & \cdot & \cdot & \cdot 11 & 1 & 1 & 11 \end{array}\right)$$

where X is a 9 x 9 integral matrix. Show that $XX^T = 10I+10J$ and $XJ=JX=10J$. By hand or computer show that no such integral matrix X exists, whence a projective plane of order 10 has no automorphism of order 11. (Hint: a row or column of X must be (2,2,2,2,2,0,0,0,0) or (3,2,2,1,1,1,0,0,0) up to order. Moreover at most one row and column of the former type can occur.)

23. Suppose that D is a symmetric (v,k,λ) design with a cyclic automorphism group G fixing a block B (setwise) and acting regularly on its points. Show that either $k=x^2+2$ for some integer x or else $k\equiv 1 \pmod 4$. (Hint: blocks other than B correspond naturally to pairs of points of B. Use this to count the number of orbits of G on blocks.)

24. Let D be a projective plane of order 34. Show that the set of primes dividing the order of the automorphism group of D is a subset of $\{3,5,7,17\}$.

pplementary Problems: Eigenvalue Techniques

These supplementary problems together prove Lemma 6, due to W. Haemers. More generally they introduce -called "eigenvalue techniques" which are quite useful in sign theory and graph theory.

We assume that the reader is familiar with elemtary facts about eigenvectors and eigenvalues typically vered in a first course in linear algebra. All matrices der discussion will have real entries. Also, throughout e rest of the text vectors are thought of as being row ctors. For these exercises only, it is convenient to spend this convention. Vectors--such as v and w--shall be lumn vectors below.

Recall that the eigenvalues of a real symmetric trix C of size n are real. We shall denote these eigenlues by

$$\lambda_1(C) \geq \cdots \geq \lambda_n(C).$$

o, note that $-\lambda_i(C) = \lambda_{n-i}(-C)$.)

. Let C be a (real) symmetric matrix of size n. For some teger i (with $0 \leq i \leq n$), let $u_1, \ldots, u_i$ be a set of orthogonal genvectors of C with eigenvalues $\lambda_1(C), \ldots, \lambda_i(C)$, respecvely. (By orthogonal, we mean with respect to the ordinary t product.) Then

(i) $\lambda_i(C) \leq \dfrac{u^T C u}{u^T u}$ for all $u \in \langle u_1, \ldots, u_i \rangle - \{0\}$.

uality holds if and only if u is an eigenvector of C for (C).

(ii) $\lambda_{i+1}(C) \geq \dfrac{u^T C u}{u^T u}$ for all $u \in \langle u_1, \ldots, u_i \rangle^{\perp} - \{0\}$.

uality holds if and only if u is an eigenvector of C for $_{+1}(C)$.

Suppose that A and B are real symmetric matrices sizes n and m, respectively. We say that the eigenvalues B interlace the eigenvalues of A if

$$\lambda_i(A) \geq \lambda_i(B) \geq \lambda_{n-m+i}(A).$$

for all $i=1,\ldots,m$. If there exists an integer k (with $0 \leq k \leq m$) such that

$$\lambda_i(A) = \lambda_i(B) \quad \text{for } i=1,\ldots,k$$

$$\text{and} \qquad \lambda_i(B) = \lambda_{n-m+i}(A) \quad \text{for } i=k+1,\ldots,m$$

the interlacing is said to be <u>tight</u>.

26. Let A be a real symmetric matrix of size n and let S be an n x m matrix such that $S^TS = I_m$. Set $B = S^TAS$. Prove:

(i) The eigenvalues of B interlace the eigenvalues of A.

(ii) If the interlacing is tight then SB=AS. (Hints: Let $u_1,\ldots,u_n$ be an orthogonal set of eigenvectors of A for $\lambda_1(A),\ldots,\lambda_n(A)$. Let $v_1,\ldots,v_m$ be an orthogonal set of eigenvectors of B for $\lambda_1(B),\ldots,\lambda_m(B)$. For $i=1,\ldots,n$, set $U_i = \langle u_1,\ldots,u_1\rangle$. For $j=1,\ldots,m$, set $V_j = \langle v_1,\ldots,v_j\rangle$ and $SV_j = \langle Sv_1,\ldots,Sv_j\rangle$. For any $1 \leq i \leq m$, choose $\tilde{w}_i$ to be a nonzero element of $SV_j \cap U_{j-1}^{\perp}$ (which has dimension at least one). Let $\tilde{v}_i$ be an element of V_j such that $\tilde{w}_i = s\tilde{v}_j$. By Problem 24,

$$\lambda_i(A) \geq \frac{\tilde{w}_i^T A \tilde{w}_i}{\tilde{w}_i^T \tilde{w}_i} = \frac{\tilde{v}_i B \tilde{v}_i}{\tilde{v}_i^T \tilde{v}_i} \geq \lambda_i(B).$$

Also

$$-\lambda_i(B) = \lambda_{m-i+1}(-B) \leq \lambda_{m-i+1}(-A) = -\lambda_{n-m+1}(A),$$

which proves (i).

Now, suppose that the interlacing is tight. For each i, equality holds in one or the other of the lines above. Hence (1) $Sv_1,\ldots,Sv_m$ are eigenvectors of A for the eigenvalues $\lambda_1(B),\ldots,\lambda_m(B)$ and (2) $v_1,\ldots,v_m$ are eigenvectors of

for these same eigenvalues. Let $V = [v_1,\ldots,v_m]$, an
x m nonsingular matrix and let $D = \text{diag}(\lambda_1(B),\ldots,\lambda_m(B))$.
en by (a) and (b) we have ASV=SVD and BV=VD. Hence
V=SBV. Multiply by the inverse of V.)

. Let A be a real symmetric matrix partitioned as follows:

$$A = \begin{bmatrix} A_{11} & \cdot & \cdot & \cdot & A_{1m} \\ \cdot & & & & \cdot \\ \cdot & & & & \cdot \\ \cdot & & & & \cdot \\ A_{m1} & & & & A_{mm} \end{bmatrix},$$

ch that A_{ii} is a square matrix of size n_i, for $i=1,\ldots,m$.
et e_{ij} be the average row sum of A_{ij} for each $i,j=1,\ldots,m$.
fine the m x m matrix $E=(e_{ij})$. Prove:

(i) The eigenvalues of E interlace the eigenvalues of A.

(ii) If the interlacing is tight then each A_{ij} has
nstant row and column sums.

Hint: Let

$$\tilde{S}^T = \begin{bmatrix} 1\ldots1 & 0\ldots0 & \ldots & 0\ldots0 \\ 0\ldots0 & 1\ldots1 & \ldots & 0\ldots0 \\ 0\ldots0 & 0\ldots0 & \ldots & 0\ldots0 \\ \cdot\quad\cdot & \cdot\quad\cdot & & \cdot\quad\cdot \\ \cdot\quad\cdot & \cdot\quad\cdot & & \cdot\quad\cdot \\ \cdot\quad\cdot & \cdot\quad\cdot & \ldots & \cdot\quad\cdot \\ 0\ldots0 & 0\ldots0 & \ldots & 1\ldots1 \end{bmatrix}$$
$$\qquad\quad \underbrace{\qquad}_{n_1} \quad \underbrace{\qquad}_{n_2} \qquad\qquad \underbrace{\qquad}_{n_m}$$

d let $D = \text{diag}(n_1,\ldots n_m)$. Set $\tilde{S}=SD^{-1}$. Then $S^TS=I_m$ and
$^T\tilde{S}=D^2$. Show that $(\tilde{S}^TA\tilde{S})_{ij}$ equals the sum of the entries
f A_{ij} and hence $E=(\tilde{S}^TA\tilde{S})D^{-2}$. Now, since $S^TAS=D^{-1}(\tilde{S}^TA\tilde{S})D^{-1}=$
^{-1}ED, the matrices S^TAS and E have the same eigenvalues
since they are similar). Apply Problem 25 to prove (i).
inally, to prove (ii) check that if $AS=S(D^{-1}ED)$ then A_{ij}
s constant row and column sums.)

28. Let N be a $m_1 \times m_2$ matrix. Set

$$A = \begin{pmatrix} 0 & N \\ N^T & 0 \end{pmatrix}.$$

Let $\lambda \neq 0$. Then the following are equivalent:

(i) λ is an eigenvalue of A of multiplicity f;

(ii) $-\lambda$ is an eigenvalue of A of multiplicity f;

(iii) λ^2 is an eigenvalue of N^TN of multiplicity f;

(iv) λ^2 is an eigenvalue of NN^T of multiplicity f;

(Hint: to show that (i) <=> (ii), let $AU=\lambda U$ for some matrix U of rank f. Write

$$U = \begin{pmatrix} U_1 \\ U_2 \end{pmatrix} \quad \text{and define} \quad \tilde{U} = \begin{pmatrix} U_1 \\ -U_2 \end{pmatrix}$$

where U_i has m_i rows (i=1,2). Then $NU_2=\lambda U_1$ and $N^TU_1=\lambda U_2$. Hence $A\tilde{U}=-\lambda\tilde{U}$. To show that (iii) <=> (iv), let $N^TNU'=\lambda U'$ for some matrix U' of rank f. Then $NN^T(NU')=\lambda NU'$. And rank NU' =rank U', since rank $\lambda U'$=rank $N^TNU' \le$ rank $NU' \le$ rank U'. Finally, to show that (i)<=> (iii), note that

$$A^2 = \begin{pmatrix} NN^T & 0 \\ 0 & N^TN \end{pmatrix}$$

and use the previous steps.)

29. Let A be the incidence matrix of a symmetric (v,k,λ) design. Set

$$A^* = \begin{pmatrix} 0 & A \\ A^T & 0 \end{pmatrix}.$$

Show that the eigenvalues of A* are $k, \underbrace{\sqrt{n},\ldots,\sqrt{n}}_{v-1 \text{ times}}, \underbrace{-\sqrt{n},\ldots,-\sqrt{n}}_{v-1 \text{ times}}, -k$

). Let D be a symmetric (v,k,λ) design with incidence
trix A. Let

$$A = \begin{pmatrix} A_1 & A_2 \\ A_3 & A_4 \end{pmatrix}$$

ere A_1 is a square matrix of size v_1. Let the average row
m of A_1 be k_1. (The average column sum is then also k_1.)
t

$$B = \begin{pmatrix} k_1 & k-k_1 \\ x & k-x \end{pmatrix},$$

ere $x = \dfrac{v_1(k-k_1)}{v-v_1}$.

(i) Observe that the entries of B are the average row
ms of the block matrices of A;

(ii) Show that $\det B = k(k_1-x)$;

(iii) Show that the eigenvalues of B are k and k_1-x;
int: find an eigenvector for k.)

. Prove Lemma 3.6. (Hint: consider the matrices

$$A^* = \begin{pmatrix} 0 & A \\ A^T & 0 \end{pmatrix} \quad \text{and} \quad B^* = \begin{pmatrix} 0 & B \\ B^T & 0 \end{pmatrix}$$

d apply Problem 26.)

The results above concerning eigenvalues can be
plied to the incidence matrix of virtually any combinatorial
ructure. In his thesis [46], Haemers uses the result
oven in Problem 26 to obtain information about strongly
gular graphs, partial geometries, chromatic numbers of
aphs, intersection numbers of block designs, and generalized
lygons. Haemers also uses the eigenvalue techniques as the
iiding principle in constructing certain new block designs,
cluding a symmetric (71,15,3) design.

NOTES TO CHAPTER 3

§3.1 Theorem 3.1 is due to Parker [110]. The variations on this theme are discussed in Dembowski [33]. See also, in this vein, Block [16]. Orbit theorems for arbitrary (not necessarily symmetric) 2-designs are discussed by Beker [14].

Theorem 3.1 and Corollary 3.2 are special cases of a lemma of Richard Brauer: if a row permutation and a column permutation have the same effect on a nonsingular matrix, the two permutations must have the same number of cycles of any given length. The matrix is intended to have entries in a field of characteristic 0. It happens that the result is true even for fields of finite characteristic, as L.G. Kovàcs shows in a letter to C.W. Curtis, which appears in Bull. London Math. Soc. 14 (1982), 127-128.

A number of authors have studied the possible automorphisms of a projective plane of order 10. Hughes [60] rejected order 2. Whitesides [140] rejected order 11. Anstee Hall and Thompson [3] rejected order 5. According to Hall [51], Janko has rejected order 3.

§3.2 For a detailed discussion of 2-transitive symmetric designs, see Kantor's articles [72] and [74].

The designs $S^{\pm}(2m)$ have turned up in many different guises. Block [16] apparently first noticed the 2-transitive automorphism group. Kantor [73] discusses these designs extensively. Our treatment follows Cameron and Seidel [28].

G. Higman [58] first constructed D_{176} by using a different approach. M. Smith [128] appears to have been the first to point out the construction based on D_{24} and the Mathieu groups. The simple group of D. Higman and C.C. Sims first arose in a different permutation representation--as a rank 3 permutation group on 100 points. For a simple introduction, see Biggs and White [15].

A peculiar property of the 2-transitive designs is at their codes (over certain fields) have remarkable low nension:

- Hamada has conjectured that the F_p-codes of the ints-hyperplane designs over F_p have minimal dimension among e F_p-codes of symmetric designs with the same parameters. e [55]. The conjecture has been checked in only a limited nber of cases. It is true for p=2; cf. Problem 8 of apter 2.
- Since H(11) is the unique symmetric (11,5,2) design s codes certainly have minimal dimension.
- The F_2-code of D_{176} has dimension 22, which seems ite low. The reader might investigate what lower bounds e can place on the dimension of the F_p-code of a symmetric 76,50,14) design. (Using the techniques in the problems to apter 2, it is not hard to show dim ≥ 14.)
- The F_2-code of the designs $S^{\pm}(2m)$ has dimension +2. (For, the subcode consisting of all sums of pairs of dewords has codimension 1 in the full code and it is the rst-order Reed-Muller code.) One can show that the F_2-code a design with the same parameters as $S^{\pm}(2m)$ has dimension least 2m+2. (Exercise.)

$S^{\pm}(2m)$ is not characterized by the dimension of its $_2$-code. We can define symmetric $(2^{2m}, 2^{2m-1} \pm 2^{m-1}, 2^{2m-2} \pm 2^{m-1})$ signs $D_i(2m)$ $(i \geq 2)$ in a similar way to the $S^{\pm}(2m)$ by placing

$$Q(x) = x_1x_2 + \ldots + x_{2m-1}x_{2m}$$

$$Q(x) = x_1x_2 + \ldots + x_{2m-1}x_{2m} + x_1x_3 \cdots x_{2i+1}$$

e F_2-code of $D_i(2m)$ also has dimension 2m+2.

Symmetric designs with parameters $(2^{2m}, 2^{2m-1} \pm 2^{m-1}, 2^{2m-2} \pm 2^{m-1})$ are in fact abundantly plentiful. Kantor [75] s shown that there are at least

$$\binom{2^n+1}{2^{n-1}} \Big/ (2^n+1)2^n(2^n-1)^2 n$$

pairwise nonisomorphic such designs. This number grows exponentially with the number of points in the design.

§3.3 The results of this section are due to Lander [79]. While in this monograph we deal only with abelian groups, the methods can be profitably applied to more complicated situations, as well. Using Proposition 3.18 and modula representation theory, Lander [81] investigates when PSL(2,q) can act as a block stabilizer of a biplane--that is, a symmetric design with $\lambda=2$.

§3.4 Contractions of incidence matrices have appeared in various guises. See Dembowski [32], Hughes [60,6 and Lander [79]. Theorem 3.15 is Dembowski [32, Satz 9(d)].

Hughes [60] proved a classical result concerning standard automorphism groups of symmetric designs. It turns out that this result provides the same number-theoretic tests applying Theorem 3.23 to groups of even order (Proposition 3.24) and groups of odd prime order (Problem 19). (Lander [79, Appendix 3] shows that the entire force of Hughes's test for groups of odd order is contained in the test for groups of od prime order.) Hence, together, Proposition 3.24 and Theorem 3.20 supersede Hughes's result.

The method of contraction can be applied to the mor general situation of a "tactical decomposition" of a symmetri design (See Dembowski [33] for a definition.) Dembowski note that Hughes's result applies to "standard" tactical decomposi tions. In this case, where the decomposition does not necessarily come from the orbit structure of a standard automorphism group, the methods of §3.3 do not apply and Hughes's result remains the strongest result known. In general, howev we are most interested in group actions.

Problems. Problem 1 is due to Baer [11]. Problem is taken from Kelly [76]. With regard to Problems 3,4,5,6, nd 7, see generally Dembowski [33]. Problem 8 is due to schbacher [10]. Problem 16 is taken from Dembowski [33]. roblem 22 was show to me by A.L. Wells. Problems 23 and 24 re due to Lander.

Supplementary Problems. These results are found in aemers's thesis [46]. See also, Haemers and Shrikhande [47].

4. DIFFERENCE SETS

§4.1 INTRODUCTION AND EXAMPLES

Let G be a group of order v, written multiplicativel A (v,k,λ)-difference set in G is a set D consisting of k group elements with the property that the list of "differences"

$$xy^{-1} \quad \text{with } x,y \in D$$

contains every nonidentity element exactly λ times. (To avoid trivialities, we insist that $k>\lambda$.)

For example, {1,2,4} is a (7,3,1)-difference set in the (additive) group of integers modulo 7, since

$1-2\equiv 6$	$2-1\equiv 1$	$4-1\equiv 3$
$1-4\equiv 4$	$2-4\equiv 5$	$4-2\equiv 2.$

Similarly, {1,3,4,5,9} is an (11,5,2)-difference set in the (additive) group of integers modulo 11. The reader will of course suspect the connection with symmetric designs:

Theorem 4.1. Let $\mathcal{D}$ be a (v,k,λ)-difference set in a group G. Define an incidence structure D, called the development of $\mathcal{D}$, as follows: the points are the elements of G and the blocks are the left translates of $\mathcal{D}$,

$$g\mathcal{D} = \{gx \mid x \in \mathcal{D}\}$$

for all $g \in G$. Then D is a symmetric (v,k,λ) design. Moreover left multiplication by G on points induces a regular automorphism group of D.

Proof. Clearly D has v points, v blocks, k points n a block and k blocks on a point. The number of points ncident with $g\mathcal{D}$ and $h\mathcal{D}$ is the number of solutions to

$$gx = hy$$

ith $x,y \in \mathcal{D}$. Since $\mathcal{D}$ is a (v,k,λ)-difference set this number s λ (for $g \neq h$). By Theorem 1.10, the structure D is a symetric (v,k,λ) design. The (left) action of G on points arries blocks to blocks and hence induces an automorphism roup regular on points and blocks.□

Conversely, if D is a symmetric (v,k,λ) design with regular automorphism group G we may view D as the developent of a (v,k,λ)-difference set in G. We identify the points f D with the elements of G as follows: choose a point x_0. or $g \in G$, identify the point gx_0 with the element g. (The hoice of which point x_0 is identified with the identity lement of G is arbitrary, but once this choice is made the dentification is completely determined.) Under this identiication, the elements of G incident with any block form a ifference set in G. We have:

Theorem 4.2. Let D be a symmetric (v,k,λ) design ith a regular automorphism group G and let x_0 be a point of . For any block B, the set

$$\mathcal{D}_B = \{g \in G \mid gx_0 \in B\}$$

s a (v,k,λ)-difference set in B. The development of $\mathcal{D}_B$ is somorphic to D.

Proof. Invert the reasoning of the previous proof.□

Difference sets and symmetric designs with regular utomorphism groups are therefore essentially the same notion. e can adopt either point of view, depending on which is more onvenient at the moment. In general, when D is a symmetric v,k,λ) design with a regular automorphism group G that is nder discussion, we shall assume that some base point x_0 has een chosen and that the points of D have been identified ith the elements of G, allowing us to freely use the language of difference sets. (It is of course crucial that the regular automorphism group of D be clearly identified since many symmetric designs admit several different regular automorphism groups, as we shall see presently.)

A difference set is <u>trivial</u> whenever the corresponding symmetric design is trivial (that is, when k=1 or k=v-1). Also, we shall use the abbreviation dev($\mathbb{D}$) for the development of the difference set $\mathbb{D}$.

Difference sets have already arisen in connection with many of the symmetric designs in previous chapters. We give some examples.

<u>Example 1</u>. The set of nonzero squares in F_q forms a $(q,\frac{1}{2}(q-1),\frac{1}{4}(q-3))$-difference set in the additive group of F_q, whenever q is a prime power congruent to 3 (mod 4). The associated symmetric design is of course H(q).

<u>Example 2</u>. Recall the connection between symmetric designs with v=4n and Hadamard matrices having constant row and column sums, introduced in Chapter 1. Suppose that D_1 and D_2 are two such symmetric designs and that H_1 and H_2 are the associated Hadamard matrices, respectively. If G_1 and G_2 are automorphism groups of D_1 and D_2, respectively, then it is immediate from the construction that $G_1 \times G_2$ acts naturally as an automorphism group of the symmetric design associated with $H_1 \otimes H_2$. In particular, if G_1 and G_2 act regularly, so does $G_1 \times G_2$.

In particular, consider the matrix

$$T = \begin{pmatrix} +1 & -1 & -1 & -1 \\ -1 & +1 & -1 & -1 \\ -1 & -1 & +1 & -1 \\ -1 & -1 & -1 & +1 \end{pmatrix}.$$

The (trivial) symmetric (4,1,0) design associated with T admits two different regular automorphism groups: $\mathbb{Z}_4$ and $\mathbb{Z}_2 \times \mathbb{Z}_2$. Thus the designs $S^{\pm}(2m)$, which are associated with $T \otimes \ldots \otimes T$ (m times), admit $(\mathbb{Z}_4)^i \times (\mathbb{Z}_2 \times \mathbb{Z}_2)^{m-i}$ as a regular automorphism group, for i=0,1,...,m. So $S^{\pm}(2m)$ gives rise to a difference set in at least m+1 different abelian 2-groups.

Example 3. The group $\mathbb{Z}_6 \times \mathbb{Z}_6$ contains a rather
egant (36,15,6)-difference set, consisting of:

(1,1)	(2,2)	(3,3)	(4,4)	(5,5)
(0,1)	(0,2)	(0,3)	(0,4)	(0,5)
(1,0)	(2,0)	(3,0)	(4,0)	(5,0) .

is difference set has v=4n and so is also connected with damard matrices having constant row and column sums. Com- ning this example with the previous, we can find difference ts in many abelian groups of order $4^a 36^b$, with parameters $n^2, 2m^2 \pm m, m^2 \pm m)$ where $m=2^{a+b-1}3^b$.

Example 4. Let us apply the construction in Theorem 12 to the affine design AG(d+1,q). The rows and columns of e parallel-class matrices $M_0,\ldots,M_r$ are indexed by the ements of AG(d+1,q), which are simply the elements of the ctor space $V=\mathbb{F}_q^{d+1}$. Since translation by any vector $v \in V$ eserves parallel classes, each of the matrices $M_0,\ldots,M_r$ left unchanged if we permute its rows and columns accord- g to translation by v (that is, if we send the column in e position indexed by the element x to the position indexed x+v, and similarly for rows). The translation group T of then acts as a regular group permuting the rows and columns $M_0,\ldots,M_r$ and preserving each of the matrices. Note that $q=p^f$ (where p is prime) then $T \simeq (\mathbb{Z}_p)^{f(d+1)}$.

Now suppose that in constructing the matrix L we ppen to use the group multiplication table of the group K, order r+1. The group K has a regular action on the rows d columns of its multiplication table, preserving the table. ecifically, for $k \in K$, send the row in the position indexed $g \in K$ to the position indexed by gk and send the column in e position indexed by $h \in K$ to the position indexed by ^{1}h. (This preserves the multiplication table since $=gkk^{-1}h$.)

The rows and columns of the incidence matrix L are dexed by ordered pairs (v,k) with $v \in V$ and $k \in K$. Combining e action of T and K, we see that T x K acts regularly on rows and columns of the matrix L, preserving it. The symmetric design can be viewed as a difference set in T x K.

So, for any prime power $q=p^f$, we can find a

$$(q^{d+1}(q^d+\ldots+q+2),\ q^d(q^d+\ldots+q+1),\ q^d(q^{d-1}+\ldots+q+1))-$$

difference set in $(\mathbb{Z}_p)^{f(d+1)} \times K$, where K is any group of order $(q^d+\ldots+q+2)$--abelian or nonabelian! For example, when q=3 and d=1, we obtain a (45,12,3)-difference set in $(\mathbb{Z}_3)^2 \times \mathbb{Z}_5$.

When p=q=2, the parameters are $(2^{2d+2}, 2^{2d+1}-2^d, 2^{2d}-2^d)$. For d=2, we obtain (16,6,2)-difference sets in $(\mathbb{Z}_2)^2 \times (\mathbb{Z}_2)^2$ and $(\mathbb{Z}_2)^2 \times \mathbb{Z}_4$. Of course, we already have such difference sets from Example 2. However for d=3, we obtain (64,28,12)-difference sets in $(\mathbb{Z}_2)^3 \times (\mathbb{Z}_2)^3$, $(\mathbb{Z}_2)^3 \times \mathbb{Z}_2 \times \mathbb{Z}_4$ and $(\mathbb{Z}_2)^3 \times \mathbb{Z}_8$. The last of these does not arise from Example 2, which only involved groups of exponent 2 or 4.

<u>Example 4½</u>. In the previous example we obtained a difference set in G = T x K. In fact, we can substantially improve upon this result. By slightly modifying the construction it is possible to obtain a difference set with the same parameters as above in any group G which contains in its center a subgroup isomorphic to T such that G/T≃K. (For the details, see Problem 8.)

To observe that this is stronger, note that $\mathbb{Z}_2 \times \mathbb{Z}_8$ has a central subgroup isomorphic to $\mathbb{Z}_2 \times \mathbb{Z}_2$ with quotient $\mathbb{Z}_4$. So, we can obtain a (16,6,2)-difference set in $\mathbb{Z}_2 \times \mathbb{Z}_8$ which did not arise from Examples 2 or 4.

More generally, we can obtain a $(2^{2t+2}, 2^{2t+1}\pm 2^{t-1}, 2^{2t}\pm 2^{t-1})$-difference set in any abelian 2-group of rank $\geq t+1$ (that is, which can be written as a product of at least t+1 cyclic groups). Example 4 produces difference sets only in abelian 2-groups of rank $\geq t+1$ in which $\mathbb{Z}_2$ appears as a direct factor at least t times. In particular, Example 4½ produces an example in a group of exponent 2^{t+2}, which Example 4 does not.

It is an open question as to which abelian 2-groups rank <t+1 possess such difference sets. The only known cessary condition, which we shall prove in §4.7 is that the oup have exponent at most 2^{t+2}.

For more on difference sets with $v=2^{2t+2}$, see oblems 5,18 and 19.

Example 5. The symmetric designs PG(m,q) always mit a regular cyclic automorphism group. (A generator is group is called a Singer cycle, after Singer [126] who und them.) To prove this, recall that points and blocks of (m,q) are the one-dimensional and m-dimensional subspaces, spectively, of the vector space $V=F_q^{m+1}$. Every F_q semi-near transformation of V induces an automorphism of PG(m,q). e easiest way to find a Singer cycle is to choose a par-cular convenient vector space V.

The field $K=F_{(q^{m+1})}$ can be viewed as an (m+1)-mensional F_q-vector space. The multiplicative group of K a cyclic group of order $q^{m+1}-1$. Let ζ be a generator of is group. Then "multiplication by ζ" defines an F_q-linear ansformation σ of K. If $x \in K-\{0\}$ then x and $\sigma^i(x)$ nerate the same one-dimensional subspace if and only if $\in F_q$, which occurs if and only if $i \equiv 0 \pmod{(q^{m+1}-1)/(q-1)}$. us $\langle x\rangle, \langle\sigma(x)\rangle, \ldots, \langle\sigma^{A-1}(x)\rangle$ are distinct one-dimensional bspaces, where $A=(q^{m+1}-1)/(q-1)$. Since A is the total mber of projective points, σ generates a regular cyclic tomorphism group of PG(m,q).

Call a difference set in a group G abelian, non-elian, cyclic, etc., if G has this property. The next ample shows that PG(m,q) gives rise to nonabelian difference ts, as well.

Example 6. Let G be the nonabelian group of order generated by two elements a and b satisfying the relations $^3=b^7=1$ and $a^{-1}ba=b^4$. Check that $\{a,a^2,b,b^2,b^4\}$ is a (21,5,1)-ifference set in G. By Problem 5 of Chapter 2, there is a ique symmetric (21,5,1) design. Thus the difference set in must give rise to the symmetric design PG(2,4).

Example 7. A great deal of work has been devoted t generalizations of the quadratic residue difference sets of Example 1.

(1) The nonzero fourth powers in $\mathbb{F}_p$ form a differen set in the additive group of the field, whenever p is a prime of the form $p=4x^2+1$, with x odd. The first example is a cycl (37,9,2)-difference set.

(2) The fourth powers including zero in $\mathbb{F}_p$ form a difference set in the additive group of the field, whenever p is a prime of the form $4x^2+9$, with x odd. The first example is a cyclic (13,4,1)-difference set (namely PG(2,3) in disguise The second is a cyclic (109,28,7)-difference set.

Sixth, eighth,...,twentieth powers have been investigated but it would take us too far afield to discuss this area of research.

Example 8. Suppose that p and p+2 are primes. Let $G = \mathbb{Z}_p \times \mathbb{Z}_{p+2}$. Let $\mathfrak{D}$ consist of all elements (a,b) such that either

(1) b=0,

(2) a and b are both nonzero squares (in their respective groups),

or (3) a and b are both nonsquares (in their respective groups).

Then $\mathfrak{D}$ forms a (4t-1, 2t-1,t-1)-difference set in G, with 4t-1=p(p+2). (For a proof, see Problem 9.) The first few examples are cyclic (15,7,3)-, (35,17,18)- and (143,71,35)-difference sets.

More generally, if q and q+2 are prime powers, the same construction supplies a difference set in the direct product of the additive groups of the fields $\mathbb{F}_q$ and $\mathbb{F}_{q+2}$. For example, we get a (63,31,15)-difference set in $(\mathbb{Z}_3)^2 \times \mathbb{Z}_7$ and a (99,49,24)-difference set in $(\mathbb{Z}_3)^2 \times \mathbb{Z}_{11}$

Suppose that $\mathfrak{D}$ is a (v,k,λ)-difference set in a group G. Certainly, $g\mathfrak{D}$ and $\mathfrak{D}h$ are also (v,k,λ)-difference sets in G for any $g,h \in G$. Also, if α is an automorphism of G, then

$$D^{\alpha} = \{\alpha(x) | x \in G\}$$

a (v,k,λ)-difference set in G. We say that a (v,k,λ)-
ference set D_1 in G is equivalent to D if

$$D_1 = gD^{\alpha},$$

some g ε G, and some automorphism α of G. Equivalent
ference sets give rise to isomorphic symmetric designs.
wever, inequivalent difference sets may give rise to
morphic designs as well.)

Sometimes it happens that, for some automorphism α,

$$D^{\alpha} = gD.$$

some g ε G. If this occurs we say that α is a multiplier
D. This has a natural interpretation in terms of the
ociated symmetric design. An automorphism α of G always
uces a point permutation of the development, dev (D);
is a multiplier if and only if this permutation induces
automorphism of dev (D).

For example, consider the difference set D={0,1,3}
$\mathbb{Z}_7$. The map $\alpha : x \longmapsto 2x$ is an automorphism of $\mathbb{Z}_7$. We
e

$$D^{\alpha} = 6 + D.$$

consider the difference set {1,3,4,5,9} in $\mathbb{Z}_{11}$. Any map
the form $\alpha : x \longmapsto bx$ where b is a nonzero square is an
omorphism of $\mathbb{Z}_{11}$ fixing the difference set.

Among all the possible permutations of G, why
uld we want to study multipliers in particular? One
swer is that there is a certain elegance to automorphisms
G which induce automorphism of dev (D). A more pragmatic
swer is that multipliers occur quite frequently. Indeed,
shall show in the next chapter that the existence of

multipliers follows immediately from simple conditions on the parameters (v,k,λ), in some cases. For example, we shall be able to show that any (73,9,1)-difference set $\mathbb{D}$ in $\mathbb{Z}_{73}$ admits the automorphism $\alpha : x \longmapsto 2x$ as a multiplier. Since α fixes a point (namely $0 \in \mathbb{Z}_{79}$) it fixes a block of dev $(\mathbb{D})$, i.e., a translate $\mathbb{D}'$ of $\mathbb{D}$. Suppose that $a \in \mathbb{D}'$. Then a, 2a, 4a, 8a, 16a, 32a, 64a, 55a, 37a are elements of $\mathbb{D}'$. Since k=9, this accounts for all of $\mathbb{D}'$. We then check that

$$\mathbb{D}' = \{a, 2a, 4a, 8a, 16a, 32a, 64a, 55a, 37a\}$$

does indeed form a difference set. Thus, starting only with the knowledge of α, we have constructed a difference set and shown that it is unique up to equivalence.

The multipliers of a difference set $\mathbb{D}$ in G form a group. Frequently we shall be concerned chiefly with automorphisms of the form $\alpha : x \longmapsto tx$ for some $t \in \mathbb{Z}$ (if we write G additively) or $\alpha : x \longmapsto x^t$ (if we use multiplicative notation). We then call α a <u>numerical automorphism</u> and, if it is a multiplier, a <u>numerical multiplier</u>. The numerical automorphisms form a subgroup contained in the center of Aut (G), automorphism group of G. Similarly, for numerical multiplier

Multipliers were originally defined only for numerical automorphisms $\alpha : x \longmapsto tx$ (whence the origin of the term multiplier). Rather than referring to α as the multiplier, the integer t was said to be the multiplier. This terminolog is firmly embedded in the literature and we too shall frequently say "t is a multiplier" of some difference set when technically we ought to say "$\alpha : x \longmapsto x^t$ is a multiplier." No confusion should arise from this abuse.

One further word about abuses. When a symmetric (v,k,λ) design D with a regular automorphism group G is under discussion we may talk about "multipliers of D." This is permissible because we have already agreed to assume that some base point x_0 of D has been chosen and the points of D identified with elements of G.

Some examples of multipliers:

(1) Let $\mathcal{D}$ be the difference set containing the non-
ro squares in F_p, with p a prime congruent to 3 (mod 4).
e additive group of F_p is $\mathbb{Z}_p$. Every automorphism of the
oup $\mathbb{Z}_p$ has the form $\alpha_t : x \mapsto tx$, for some integer $t \neq 0$
od p). The automorphism α_t is a multiplier of $\mathcal{D}$ if and
ly if t is a quadratic residue (mod p). Concerning the
fference set of nonzero squares in F_q, see Problem 11.

(2) Recall Example 2. Suppose that G_1 and G_2 act
ularly on D_1 and D_2 respectively. If α_1 and α_2 are multi-
iers of D_1 and D_2 respectively, then the automorphism
$\times \alpha_2$ of $G_1 \times G_2$ is a multiplier of the symmetric design
ained by the Kronecker product construction. For example,
e (4,1,0)-difference set in either $\mathbb{Z}_4$ or $\mathbb{Z}_2 \times \mathbb{Z}_2$ has the
$x \mapsto -x$ as a multiplier. Thus, so do all the difference
ts obtained by taking Kronecker products of these. So, by
r abuse of terminology, we say that -1 is a multiplier of
ese difference sets.

Note also that -1 is a multiplier of the (36,15,6)-
fference set in $\mathbb{Z}_6 \times \mathbb{Z}_6$, given in Example 3.

(3) In Example 5, we defined PG(m,q) by using
F_q^{m+1} as our F_q-vector space. "Multiplication by ζ" pro-
ded an automorphism σ generating a regular cyclic group.

Suppose that $q=p^f$ (with p prime) and let α be the
eld automorphism of K such that $\alpha(\zeta)=\zeta^p$. For $x \in K$, we have

$$\alpha\sigma^i(x) = \sigma^{ip}(x)$$

us α acts as an automorphism of the group $\langle\sigma\rangle$ sending σ to
. Moreover α defines an F_q-semilinear transformation of K
d thus induces an automorphism of PG(m,q). Hence, α is a
ltiplier! Or, in terms of our abuse of terminology,
is a multiplier of the cyclic difference set associated
th PG(m,q).

(4) The points of $S^{\pm}(2m)$ are the elements of the
ctor space $V=F_2^{2m}$, which is an elementary abelian 2-group.
the construction, translation by an element of V induces
automorphism of $S^{\pm}(2m)$. So $S^{\pm}(2m)$ admits the regular
oup V. The automorphisms of this group are simply the

nonsingular linear transformations of the vector space V. The linear transformations preserving $S^{\pm}(2m)$ are just those preserving the bilinear form--namely, the elements of Sp(2m,2). Thus Sp(2m,2) acts as the full multiplier group of $S^{\pm}(2m)$. Notice that the difference set has the extraordinary property that the multipliers act transitively on nonidentity elements of the group. (We explore this further in §4.2.)

It is sometimes useful to express properties of difference sets in a group G in terms of the integral group ring $\mathbb{Z}G$. Some notation: If S is a subset of G, we write $\underline{S}$ for the element

$$\underline{S} = \sum_{g \in S} g$$

in $\mathbb{Z}G$. Also, if $x = \sum_{g\in G} a_g g$ is an element of $\mathbb{Z}G$, we write $x^{(t)}$ for the element

$$x^{(t)} = \sum_{g\in G} a_g g^t$$

in $\mathbb{Z}G$.

<u>Proposition 4.3</u>. <u>A subset</u> $\mathcal{D}$ <u>of</u> G <u>is a</u> (v,k,λ)-<u>difference set if and only if</u>

$$\underline{\mathcal{D}}\,\underline{\mathcal{D}}^{(-1)} = (k-\lambda)\underline{1}_G + \lambda\underline{G}.$$

<u>Proof</u>. Immediate from the definitions.□

If α is an automorphism of G and $x = \sum a_g g$ is an element of $\mathbb{Z}G$, we write x^{α} for the element

$$x^{\alpha} = \sum a_g \alpha(g).$$

Thus α is a multiplier of the difference set $\mathcal{D}$ if and only if $\underline{\mathcal{D}}^{\alpha} = g\underline{\mathcal{D}}$, for some $g \in G$. The group ring is useful not simply because we can use it to express properties so succinctly. We will find later that it is a helpful algebraic tool for studying difference sets.

Some important questions we should like to answer are: for which (v,k,λ) does there exist a (v,k,λ)-difference set? an abelian (v,k,λ)-difference set? a cyclic (v,k,λ)-difference set? Of course, (v,k,λ) must satisfy the necessary conditions for the existence of a symmetric (v,k,λ) design. However we should expect further requirements since we are demanding the existence of a large automorphism group. While in this section we have concentrated on constructions of difference sets, in subsequent sections we shall focus on requirements that lead to nonexistence theorems. Our first such result comes from a straightforward application of the work of §3.3.

Theorem 4.4. Suppose that there exists a (v,k,λ)-difference set $\mathcal{D}$ in a group G. If for some divisor w of v, greater than 1, and some prime p there exists an integer j such that

$$p^j \equiv -1 \pmod{w}$$

then p does not divide the square-free part of $n=k-\lambda$.

Proof. Suppose that p divides n. There is no harm in assuming that w is prime (if not, replace it by a prime divisor of w) and that $w \neq p$. Consider the symmetric design $D = \text{dev}(\mathcal{D})$. Let H be a subgroup of G having order w. Then H acts as a standard automorphism group of D fixing zero (an even number of) points. By Theorem 3.20, if $p^j \equiv -1 \pmod{w}$ for some integer j, then p cannot divide n^*, the square-free part of n. $\square$

We can use Theorem 4.4 to exclude possible parameters of difference sets. For example, there cannot exist a difference set with parameters (155,22,3) since $19 \equiv -1 \pmod{5}$ or one with parameters (55,27,13) since $2^2 \equiv -1 \pmod{5}$. Table 4-1 shows the parameters (v,k,λ) with $k \leq 50$ excluded by this test.

Table 4-1

(v,k,λ)	$p^f \equiv -1 \pmod{w}$
(25,9,3)	$2^2 \equiv -1 \pmod{5}$
(111,11,1)	$2 \equiv -1 \pmod{3}$
(157,13,1)	$3^{39} \equiv -1 \pmod{157}$
(241,16,1)	$3^{60} \equiv -1 \pmod{241}$
(121,16,2)	$2^{55} \equiv -1 \pmod{121}$
(61,16,4)	$3^5 \equiv -1 \pmod{61}$
(41,16,6)	$2^{10} \equiv -1 \pmod{41}$
(39,19,9)	$2^6 \equiv -1 \pmod{13}$
(421,21,1)	$5^{105} \equiv -1 \pmod{421}$
(155,22,3)	$19 \equiv -1 \pmod{5}$
(201,25,3)	$2 \equiv -1 \pmod{3}$
(61,25,10)	$3^5 \equiv -1 \pmod{61}$
(51,25,12)	$13^2 \equiv -1 \pmod{17}$
(703,27,1)	$2^9 \equiv -1 \pmod{19}$
(55,27,13)	$2^2 \equiv -1 \pmod{5}$
(85,28,9)	$19 \equiv -1 \pmod{5}$
(407,29,2)	$3^9 \equiv -1 \pmod{37}$
(291,30,3)	$3^{24} \equiv -1 \pmod{97}$
(265,33,4)	$29 \equiv -1 \pmod{5}$
(97,33,11)	$2^{24} \equiv -1 \pmod{97}$
(1191,35,1)	$2 \equiv -1 \pmod{3}$
(1261,36,1)	$5^2 \equiv -1 \pmod{13}$
(85,36,15)	$7^2 \equiv -1 \pmod{5}$
(149,37,9)	$7^{37} \equiv -1 \pmod{149}$
(75,37,18)	$19 \equiv -1 \pmod{5}$
(1561,40,1)	$3^3 \equiv -1 \pmod{7}$

Table 4-1 (continued)

(v,k,λ)	$p^f \equiv -1$	(mod w)
(521,40,3)	$37^{65} \equiv -1$	(mod 521)
(1641,41,1)	$2 \equiv -1$	(mod 3)
(411,41,4)	$37^2 \equiv -1$	(mod 137)
(575,42,3)	$3^2 \equiv -1$	(mod 5)
(87,43,21)	$2 \equiv -1$	(mod 3)
(199,45,10)	$5^{99} \equiv -1$	(mod 199)
(91,45,22)	$23^3 \equiv -1$	(mod 13)
(95,47,23)	$3^2 \equiv -1$	(mod 5)
(1129,48,2)	$2^{282} \equiv -1$	(mod 1129)
(565,48,4)	$11^{27} \equiv -1$	(mod 113)
(785,49,3)	$3^{84} \equiv -1$	(mod 337)
(169,49,14)	$5^2 \equiv -1$	(mod 13)
(113,49,21)	$7^7 \equiv -1$	(mod 113)
(337,49,7)	$3^{84} \equiv -1$	(mod 337)
(351,50,7)	$43^3 \equiv -1$	(mod 13)

Remark. Theorem 4.4 has long been known for abelian ifference sets. The proof, however, is the first to include onabelian difference sets as well.

In the next sections we shall show that by restrict- ng attention to abelian difference sets we can obtain many tronger existence criteria.

§4.2. ABELIAN DIFFERENCE SETS

For abelian difference sets, we have the following theorem.

Theorem 4.5. *Suppose that there exists a* (v,k,λ)-*difference set* $\mathcal{D}$ *in an abelian group* G. *If for some divisor* w *of* v *greater than* 1 *and some prime* p, *there exists an integer* j *such that*

$$p^j \equiv -1 \pmod{w}$$

then p *does not divide the squarefree part of* $n=k-\lambda$.

Moreover, if w *is the exponent of* G, *then* p *does not divide* n.

Proof. The first part of the theorem is simply Theorem 4.4 which we repeat for completeness. Now, let w be the exponent of G and let p be a prime such that $p^j \equiv -1$ (mod w) for some integer j. Suppose that p divides n. After possibly replacing $\mathcal{D}$ by its complement we may assume that p divides λ (since $\lambda\lambda'=n(n-1)$).

Let C be the $\mathbb{F}_p$-code of dev ($\mathcal{D}$). As we observed in Chapter 2, the code C is a self-orthogonal code (with respect to the dot product) in $V=\mathbb{F}_p^{\,v}$. Because $\mathcal{D}$ is a difference set, V and C acquire the structure of $\mathbb{F}_pG$-modules. Indeed $V\simeq\mathbb{F}_pG$.

By Appendix D the condition that $p^j \equiv -1$ (mod w) for some j implies that every irreducible $\mathbb{F}_pG$-module is self-contragredient. By the remark following the proof of Proposition 3.18 it follows that every composition factor of C occurs as a composition factor of V with even multiplicity. However, we know from Appendix D that every composition factor of an abelian group ring $\mathbb{F}_pG$ occurs only once (when p does not divide $|G|$). We conclude that C has no composition factor at all. That is, C is the zero code. This is impossible since dim $C\geq 2$ by Proposition 2.6. We have reached this contradiction by assuming that $p|n$. Hence $p\nmid n$.□

Theorem 4.5 excludes many cases that pass the test of Theorem 4.4. For example, there does not exist a (343,19,1)-difference set in an abelian group since $3^{147} \equiv -1$ (mod 343). The table below shows which possibilities are excluded with k≤50. (The third column gives the structure of G as a product of cyclic groups.)

(v,k,λ)	n	Group	$p^f \equiv -1$	(mod w)
(56,11,2)	9	(2)(4)(7)	$3^3 \equiv -1$	(mod 28)
		$(2)^3(7)$	$3^3 \equiv -1$	(mod 14)
(40,13,4)	9	$(2)^3(5)$	$3^2 \equiv -1$	(mod 10)
(343,19,1)	18	(343)	$3^{147} \equiv -1$	(mod 343)
		(7)(49)	$3^{21} \equiv -1$	(mod 49)
		$(7)^3$	$3^3 \equiv -1$	(mod 7)
(139,24,4)	20	(139)	$2^{69} \equiv -1$	(mod 139)
(121,25,5)	20	121	$2^{55} \equiv -1$	(mod 55)
		$(11)^2$	$2^5 \equiv -1$	(mod 11)
(79,27,9)	18	(79)	$3^{39} \equiv -1$	(mod 79)
(249,32,4)	28	(3)(83)	$2^{41} \equiv -1$	(mod 249)
(171,35,7)	28	(9)(19)	$2^9 \equiv -1$	(mod 171)
		$(3)^2(19)$	$2^9 \equiv -1$	(mod 57)
(112,37,12)	25	(2)(61)	$5^{15} \equiv -1$	(mod 112)
(131,40,12)	28	(131)	$2^{65} \equiv -1$	(mod 131)
(259,43,7)	36	(7)(37)	$3^9 \equiv -1$	(mod 259)
(589,49,4)	45	(19)(31)	$3^{45} \equiv -1$	(mod 589)

Let us elaborate on the proof of Theorem 4.5, which contains the key idea which we shall use over and over again in this chapter. Suppose that G is an abelian group of order v. Consider the vector space $V=F_p^v$, on which we let G act by regularly permuting the coordinates. Define a G-<u>code</u> to be a subspace of V closed under the action of G.

Clearly V is isomorphic (as a G-module) to the group ring $\mathbb{F}_pG$; the G-codes are simply the ideals, or sub-$\mathbb{F}_pG$-modules. (In view of this correspondence, we shall use the notation $\sum_{g\in G} a_g g$ for elements of V, whenever it is more convenient or clearer than using v-tuples.)

Let α be an automorphism of G. Then α acts naturally on V. If $x = \sum a_g g$ then $\alpha(x) = \sum a_g \alpha(g)$. We say that α is a *multiplier* of the $\mathbb{F}_pG$-code C whenever

$$C = \alpha(C) = \{\alpha(x)\,|\,x\varepsilon C\}.$$

Clearly, α is a multiplier of C if and only if, when considered as an ideal of $\mathbb{F}_pG$, the code C equals its own twist by α. In view of the work in Appendix D, we have the following result.

Proposition 4.6. Let G be an abelian group of order v and let p be a prime not dividing v. An automorphism α of G is a multiplier of a G-code C defined over $\mathbb{F}_p$ if and only if the set of (absolutely irreducible) characters involved in the module C is closed under twisting by α.

Let C be the $\mathbb{F}_p$-code of a difference set in G. Then C is certainly a G-code. Moreover, if α is a multiplier of the difference set it is *a fortiori* a multiplier of the G-code. The converse, however, is false. The G-code will in general possess other multipliers, aside from those inherited from the difference set. In fact, C will always have the automorphism $\alpha\colon g \longmapsto g^p$ as a multiplier. For, by Proposition D.13, we have the following result.

Proposition 4.7. Let G be an abelian group of order v, let p be a prime not dividing v and let α be the automorphism of G given by $\alpha\colon g \longmapsto g^p$.

Then α is a multiplier of every G-code. Or, to abuse terminology, p is a multiplier of every G-code.

Remark. If we did not need representation theory around anyway for other purposes, we would prefer this simpler proof of Proposition 4.7: Let $u = \sum a_g g \in C$. Since C is an ideal, $u^p \in C$. But then

$$u^p = (\sum a_g g)^p \equiv \sum a_g{}^p g^p \equiv \sum a_g g^p = u^{(p)} \pmod{p}.$$

Hence $u^{(p)} \in C$ and C is fixed under the map $\alpha: g \longmapsto g^p$ for $x \in G$.

A very important automorphism of G is the map $\beta: g \longmapsto g^{-1}$. Call a G-code reversible if β is a multiplier. Recall from Appendix D that twisting a character by β, replaces the character by its contragredient.

Proposition 4.8. *Let G be an abelian group of order v and let p be a prime not dividing v. Suppose that a G-code C in F_pG is both reversible and self-orthogonal with respect to the ordinary dot product. Then C=(0).*

Proof. We use the same idea as in the proof of Proposition 3.12. We have

$$0 \subseteq C \subseteq C^{\perp} \subseteq V = F_pG .$$

Suppose that C is not the zero code. Let N be a composition factor of C. Since C is reversible, then N^{β} is also a composition factor of C. But $N^{\beta} \simeq N^*$, the contragredient. Thus, N and N* are composition factors of C. However, by the proof of Proposition 3.12, both N and N* must be composition factors of $V/C^{\perp}$. Hence, N occurs as a composition factor of $V=F_pG$ with multiplicity at least two. This contradicts the fact, proven in Appendix D, that every composition factor of F_pG occurs with multiplicity one. Hence C must be the zero code.□

Proposition 4.8 is the central tool in this chapter, which we can apply in numerous situations. For example, we can interpret the proof of Theorem 4.5 in terms of the point of view just developed. Let $\mathcal{D}$ be a (v,k,λ)-difference set in an abelian group G. Suppose that p is a prime not dividing v and let C be the F_p-code of dev $(\mathcal{D})$. By Proposition 4.7, the automorphism $\alpha^j: g \longmapsto g^{p^j}$ is a multiplier of C, for every integer j. If $p^f \equiv -1$ (mod exponent of G) for some integer f, then C is reversible. Since C is not the zero code, it cannot be self-orthogonal. Thus, p cannot divide n.

The next three sections consist essentially of locating diverse situations in which Proposition 4.8 applies. Before turning to this task, we digress to mention how Proposition 4.8 can be used to help determine the dimension of a G-code. We begin with a lemma.

Lemma 4.9. Let G be an abelian group and let $G^\times$ be the group of absolutely irreducible characters of G (with values in a field K of characteristic not dividing $|G|$). Any automorphism α fixes as many elements of G as characters in $G^\times$.

Proof. Let C be the character table of G, which is an invertible matrix since the characteristic of K does not divide $|G|$. By the definition of the action of α on $G^\times$, permuting rows of C according to the action of α on elements of G has the same effect as permuting columns of C according to the action of α on characters. Arguing just as in Proposition 3.1, the number of fixed rows and columns must be equal. □

The lemma has certain immediate consequences. A group of automorphisms has as many orbits on elements of G as on characters in $G^\times$. An automorphism has the same cycle structure on elements as on characters. An automorphism group is transitive on nonidentity elements if and only if it is transitive on nonprincipal characters.

Theorem 4.10. Let D be the symmetric $(p,\frac{1}{2}(p-1),\frac{1}{4}(p-3))$ design H(p), where p is a prime congruent to 3 (mod 4). Let r be a prime dividing the order n of this design and let C be the F_r-code of D.

The dimension of C is $\frac{1}{2}(p+1)$. The extended F_r-code of D is self-dual (with respect to $\psi(x,y) = x_1y_1 + \ldots + x_py_p - \lambda x_{p+1}y_{p+1}$).

Proof. A cyclic group G of order p acts regularly on D, so we may view C as an ideal in F_rG. We shall investigate how many algebraically irreducible characters (defined ove an extension field K containing p-th roots of unity) are involved in C. This number is precisely the dimension of C.

The nonzero codeword (k,...,k) lies in C, since it
s the sum of the rows of the incidence matrix of D. Thus,
contains the unique nonzero submodule on which G acts
ivially. That is, C involves the principal character χ_0.
ince dim C$\geq$2, it also involves at least one nonprincipal
aracter χ.

Now the multiplier group M of D is cyclic and has
bits of size $\frac{1}{2}$(p-1) on nonidentity elements of G. Thus,
has two orbits of size $\frac{1}{2}$(p-1) on nonprincipal characters.
ince M is a group of multipliers of C, then C must involve
l characters in the same orbit as χ. Hence C involves at
ast 1 + $\frac{1}{2}$(p-1) = $\frac{1}{2}$(p+1) characters and dim C $\geq \frac{1}{2}$(p+1). By
oposition 2.6, we have dim C = $\frac{1}{2}$(p+1).

Since the extended F_r-code also has dimension $\frac{1}{2}$(p+1)
d is self-orthogonal with respect to ψ it is self-dual.□

Consider, in particular, D=H(383). The order n of
e design is 96. Since $2^5 \| n$, we can produce a sequence of
-codes,

$$0 \subseteq C_0 \subseteq C_1 \subseteq C_2 \subseteq C_3 \subseteq C_4 \subseteq F_2^{384},$$

the method of Section 2.3. By Theorem 4.10, the dimension
C_0 is $\frac{1}{2}$(384) = 192. Hence $C_0=C_1=C_2=C_3=C_4$. This shows that
e codes in the chain need not be distinct, answering a
estion raised in Problem 22 of Chapter 2.

Remark. The code C is called a quadratic residue
de and it is very important in algebraic coding theory.
e restriction that p be a prime is inessential; prime powers
rk as well. (See Problem 11.)

Another interesting application is the following.

Theorem 4.11. *Suppose that* $\mathcal{D}$ *is a* (v,k,λ)-*difference*
t in a group G. *If the multipliers of* $\mathcal{D}$ *act transitively*
the nonidentity elements of G *then*

$$(v,k,\lambda) = (2^{2m}, 2^{2m-1} \pm 2^{m-1}, 2^{2m-2} \pm 2^{m-1})$$

r some integer m *and* G *is an elementary abelian* 2-*group*.

<u>Proof</u>. Let G be a group with a group A of automorphisms acting transitively on nonidentity elements. Any automorphism preserves the order of any element. Hence all nonidentity elements have the same order, which must be a prime p. Now A preserves the center of G, whence the center is either {1} or G. Since a p-group never has a trivial center, the center of G must be G. Hence G is an abelian p-group.

Now, suppose that $\mathcal{D}$ is a (v,k,λ)-difference set in G and that A is its group of multipliers. Suppose that some prime r, other than p, divides the order n. Let C be the $\mathbb{F}_r$-code of the development of D. The group A is a multiplier group of C. Since r does not divide $|G|$, Lemma 4.9 applies and A must act transitively on nonprincipal (absolutely irreducible) characters of G. Arguing as in the previous theorem we see that C either involves every nonprincipal character or none of them. Hence dim $C \leq 1$ or dim $C \geq v-1$. But this contradicts Proposition 4.6. We conclude that no prime other than p can divide n.

Now, by Problems 3 and 4 of Chapter 2, if v and n are powers of a prime p, then we must have p=2 and

$$(v,k,\lambda) = (2^{2m}, 2^{2m-1} \pm 2^{m-1}, 2^{2m-2} \pm 2^{m-1})$$

for some integer m.□

<u>Corollary 4.12</u>. <u>Let D be a symmetric</u> (v,k,λ) <u>design with a doubly transitive automorphism group</u> G. <u>If</u> G <u>contains a regular normal subgroup</u> T, <u>then</u>

$$(v,k,\lambda) = (2^{2m}, 2^{2m-1} \pm 2^{m-1}, 2^{2m-2} \pm 2^{m-1})$$

<u>for some integer</u> m, <u>and</u> T <u>must be an elementary abelian</u> 2-<u>group</u>.

<u>Proof</u>. The proof consists of noting that the corollary describes the same situation as the preceding theorem. Fix a point x of D and consider the group G_x, stabilizing x. Because G is 2-transitive, G_x is transitive

on the points other than x. For every $\sigma \in G_x$, conjugation by σ defines an automorphism of T (since T is normal in G). If t is the unique element of T carrying x to some point y then $\sigma t \sigma^{-1}$ is the unique element of T carrying x to σy. Hence, the group of automorphism of T defined by conjugation by elements of G_x is transitive on nonidentity elements of T.

So, we interpret D as the development of a difference set in T and G_x as a group of multipliers satisfying the conditions of Theorem 4.11.□

<u>Remark</u>. If we add the condition that for some block B, the group G_B acts 2-transitively on the points on B and off B then D must be $S^{\pm}(2m)$, by Theorem 3.14. It seems likely that the result holds without this added condition, but no proof is known (without invoking the classification of 2-transitive groups).

§4.3. CONTRACTING DIFFERENCE SETS

In the next six sections we shall prove a variety of nonexistence results for abelian difference sets. While we shall employ a number of different approaches--some based on simple counting arguments, others exploiting the structure of KG-modules and still others relying on algebraic number theory--all the results are essentially variations on a single theme: that the reversibility of any code associated with a difference set places strong restrictions on the difference set.

In order to prove theorems about difference sets in an abelian group G, it is important to be able to focus attention on subgroups and quotient groups of G. The best way to do this is the method of contraction, introduced in §3.4.

Suppose that $\mathbb{D}$ is a (v,k,λ)-difference set in an abelian group G. Let D be the development of $\mathbb{D}$ and let A be the incidence matrix of D. We can contract A by any subgroup H of G. The orbits of H correspond to cosets of G in H; there are $w := [G:H]$ orbits of size $h := |H|$. As we observed in §3.4, the contracted matrix A_H has size w and satisfies the equations

$$A_H A_H^T = nI + h\lambda J$$

and

$$A_H J = J A_H = kJ.$$

Just as A defines an incidence structure on the elements of G, it is helpful to pretend that A_H defines an incidence structure on the elements of G/H. I say "pretend" because in general the entries of A_H will not be only 1 and 0. We could of course formally define structures with multiple incidence, but that is unnecessary. Rather, we simply carry over the language of symmetric designs in the obvious way. View columns of A_H as referring to "points", and rows as referring to "blocks" of some "contracted incidence structure," which we might as well call D_H. We define "automorphisms" of D_H to be permutations of the points and blocks which preserve A_H. It should not be hard to see that the group G/H, which acts naturally on the points and blocks of D_H, induces an automorphism group. (This is true in the nonabelian case as well, provided that H is normal in G.) So, we may view D_H as the "development of a contracted difference set" $\mathcal{D}_H$ in G/H. For example the matrix

$$A = \begin{pmatrix}
0&0&0&1&0&0&1&1&0&0&0&0&1&0&1&0&0&0&0&0&0\\
0&0&0&0&1&0&0&1&1&0&0&0&0&1&0&1&0&0&0&0&0\\
0&0&0&0&0&1&0&0&1&1&0&0&0&0&1&0&1&0&0&0&0\\
0&0&0&0&0&0&1&0&0&1&1&0&0&0&0&1&0&1&0&0&0\\
0&0&0&0&0&0&0&1&0&0&1&1&0&0&0&0&1&0&1&0&0\\
0&0&0&0&0&0&0&0&1&0&0&1&1&0&0&0&0&1&0&1&0\\
0&0&0&0&0&0&0&0&0&1&0&0&1&1&0&0&0&0&1&0&1\\
1&0&0&0&0&0&0&0&0&0&1&0&0&1&1&0&0&0&0&1&0\\
0&1&0&0&0&0&0&0&0&0&0&1&0&0&1&1&0&0&0&0&1\\
1&0&1&0&0&0&0&0&0&0&0&0&1&0&0&1&1&0&0&0&0\\
0&1&0&1&0&0&0&0&0&0&0&0&0&1&0&0&1&1&0&0&0\\
0&0&1&0&1&0&0&0&0&0&0&0&0&0&1&0&0&1&1&0&0\\
0&0&0&1&0&1&0&0&0&0&0&0&0&0&0&1&0&0&1&1&0\\
0&0&0&0&1&0&1&0&0&0&0&0&0&0&0&0&1&0&0&1&1\\
1&0&0&0&0&1&0&1&0&0&0&0&0&0&0&0&0&1&0&0&1\\
1&1&0&0&0&0&1&0&1&0&0&0&0&0&0&0&0&0&1&0&0\\
0&1&1&0&0&0&0&1&0&1&0&0&0&0&0&0&0&0&0&1&0\\
0&0&1&1&0&0&0&0&1&0&1&0&0&0&0&0&0&0&0&0&1\\
1&0&0&1&1&0&0&0&0&1&0&1&0&0&0&0&0&0&0&0&0\\
0&1&0&0&1&1&0&0&0&0&1&0&1&0&0&0&0&0&0&0&0\\
0&0&1&0&0&1&1&0&0&0&0&1&0&1&0&0&0&0&0&0&0
\end{pmatrix}$$

the incidence matrix of a symmetric (21,5,1) design D, itten in such a way as to make clear a cyclic regular auto-rphism group. Let H be the subgroup of order 3 in G. Then

$$A_H = \begin{pmatrix} 2&0&0&1&0&1&1 \\ 1&2&0&0&1&0&1 \\ 1&1&2&0&0&1&0 \\ 0&1&1&2&0&0&1 \\ 1&0&1&1&2&0&0 \\ 0&1&0&1&1&2&0 \\ 0&0&1&0&1&1&2 \end{pmatrix}.$$

e cyclic group G/H of order 7 acts on A_H.

We continue this legal fiction by defining multi-iers. If an automorphism α of G/H induces, by its action points, an automorphism of D_H, we say that α is a "G/H-ltiplier." Just as D_H inherits its regular automorphism oup from D it also inherits certain multipliers. If α is multiplier of D and $\alpha(H)=H$, then α induces an automorphism G/H which is a G/H-multiplier of D_H. In particular, this true for every numerical automorphism. In the example ove we see that 2 is a multiplier of D and a G/H-multiplier D_H.

Most of the results we have obtained for symmetric signs have depended essentially on the equations $AA^T=nI+\lambda J$ d $AJ=JA=kJ$. We can immediately obtain analogues for the ntracted incidence structure D_H by using the analogous uations.

For example, all of the results of §3.1 go through th no trouble. Similarly--defining "semi-standard" and "stan-rd" automorphism groups of D_H in the obvious way--we can ove contracted versions of Theorem 3.21 and Proposition 3.22. he reader should state these contracted versions.) Using e latter result, we can prove the following theorem about elian difference sets.

<u>Theorem 4.13</u>. <u>Let $\mathcal{D}$ be a (v,k,λ)-difference set in abelian group G. Let H be a subgroup of G. If the integer is a G/H-multiplier then either</u>

(1) n is a square, or

(2) t is a square (mod w), where $w = |G/H|$.

In particular, if t is a multiplier of $\mathcal{D}$, then either n is a square or t is a square (mod v).

Proof. Let us suppose that n is not a square (otherwise there is nothing to prove) and that t is a G/H-multiplier. Let p be a prime dividing w. We can choose a subgroup K such that $H \subseteq K \subseteq G$ and $|G/K| = p$. Now, t is also a G/K-multiplier. By applying the contracted version of Proposition 3.17 to the group of G/K-multiplier generated by t, we see that t must have an odd number of orbits on the elements of G/K which is a cyclic group of order p. The automorphism $\alpha: g \longmapsto g^t$ has an odd number of orbits on this group if and only if t is a square (mod p). (Check this.) Thus, for every prime p dividing w, we see that t is a square (mod p).

Notice that w must be odd (for otherwise v would be even and n a square). It is a matter of elementary number theory that for any odd integer w, an integer t is a square (mod w) if and only if t is a square (mod p) for every prime p dividing w. (See Problem 14.) Finally, the last statement follows by taking H to be the identity subgroup.□

Let us carry on with the analogies. As we noted in §3.4, we can use the contracted incidence matrix to produce codes. Whenever p is a prime dividing n*, we can obtain an $\mathbb{F}_p$-code of length w+1, self-dual with respect to an appropriate scalar product. (This follows from Theorem 3.23 if p does not divide |H| and, in general, from the second remark following Theorem 3.23.) These self-dual codes admit all automorphisms of D_H and we can use our standard methods on them. A slight modification of Theorem 4.15 yields the following result.

Theorem 4.14. Suppose that there exists a (v,k,λ)-difference set $\mathcal{D}$ in an abelian group G. Suppose that H is a proper subgroup of G and that the integer t is a G/H-multiplier. Let p be a prime not dividing the order of G/H. If there exists an integer j such that

$$tp^j \equiv -1 \pmod{\text{exponent of } G/H}$$

en p does not divide the square-free part of n.

Moreover, if H is the identity subgroup, then p es not divide n.

Proof. Suppose that the hypotheses hold but that p vides n^*. Let $w=|G/H|$. We can produce from D_H an $\mathbb{F}_p$-code of length w+1 which is self-dual with respect to an appro- iate scalar product. Let E be the subcode of C consisting all codewords with a 0 in the (w+1)-st coordinate. The nension of E is at least $\frac{1}{2}(w-1)$. Dropping the last coordi- te of E, we obtain a G-code that is self-orthogonal with spect to the dot product.

Now, t is a multiplier of E (by the construction of and p is also a multiplier (by Proposition 4.7). Thus ^{f} is a multiplier and E is reversible. Thus E is the zero de and $\frac{1}{2}(w-1)=\dim E=0$. But then w=1, contradicting the pothesis that H is a proper subgroup. Hence $p\nmid n^*$.

The "moreoever" part is proven similarly, by using e ordinary extended $\mathbb{F}_p$-code of D (just as in Theorem 4.5).□

Corollary 4.15. Let D be a (v,k,λ)-difference set an abelian group G. Suppose that -1 is a G/H-multiplier, r some proper subgroup H. Then either n is a square, or r some prime p,

(1) the square-free part of n is p

(2) the order of G/H is a power of p and

(3) $p\equiv 1 \pmod 4$.

Proof. Suppose that n is not a square. Let p be prime dividing the square-free part of n. If $|G/H|$ is not power of p then we can choose a subgroup J such that $\leq J \leq G$ and $|G/J|$ is relatively prime to p. Since -1 is then G/J-multiplier, Theorem 4.14 is then violated. Thus $|G/H|$ a power of p. No other prime r can divide n^*, for then $/H|$ would have to be a power of r. Hence $n^* = p$. Finally, eorem 4.13 shows that -1 must be a square (mod p). Hence, $\equiv 1 \pmod 4$.□

Of course, the corollary is only useful as a tool for excluding difference sets once we know that -1 must actually be a multiplier of some particular (v,k,λ)-differenc set. In Chapter 5, we will prove certain theorems along thes lines. For example, we will be able to show that 2 must be a multiplier of any abelian (813,29,1)-difference set. If G is the group of order 271, and H is a subgroup of order 271, then 2 is also a G/H-multiplier. Since $|G/H| = 3$ and $2 \equiv -1$ (mod 3), the corollary is violated. We must conclude that no abelian (813,29,1)-difference set exists.

So far we have not been able to say much about the case in which n is a square. In the next section we develop methods which work in this case.

§4.4 G-MATRICES

In this section we explore the properties of the incidence matrix of a difference set and prove some elementar relations which must hold among the various parameters.

We begin with the notion of a G-matrix, complementin the idea of G-codes. Let G be a finite abelian group with elements $g_1,\ldots,g_w$, where g_1 is the identity element. A G-*matrix* is a w x w matrix A such that if

$$(a_{g_1},\ldots,a_{g_w})$$

is the first row of A then the jth row of A is

$$(b_{g_1},\ldots,b_{g_w}),$$

where

$$b_{g_i} = a_{g_j^{-1}g_i}.$$

In other words, the columns correspond to elements of G and the jth row is obtained by letting g_j act by left multiplication on the first row. (Note that for the purpose of discussing G-matrices we choose a fixed order in which to list the elements.) The F_p-span of the rows of an integral G-matri is a G-code.

G-matrices can also be described easily using the left regular representation $\bar{\rho}$ of G. The G-matrices are just the linear combinations of the permutation matrices $\bar{\rho}(g)$, for $g \in G$. (Incidentally this shows that the product of G-matrices is again a G-matrix.)

Since we will frequently use the condition, we define a prime p to be *semiprimitive* (mod w) if $p^j \equiv -1$ (mod w) for some integer j. More generally, say that an integer m is *semiprimitive* (mod w) if every prime factor of m is semiprimitive (mod w). (Necessarily, $(m,w)=1$.)

Lemma 4.16. Let G *be an abelian group of order* w *and suppose that* U *is an integral* G-*matrix such that*

$$UU^T \equiv 0 \pmod{m^2}$$

for some integer m. *If* m *is semiprimitive* (modulo exponent *of* G), *then*

$$U \equiv 0 \pmod{m}.$$

Proof. We proceed by induction on the number of positive divisors of m. If m has one divisor then m=1, in which case the conclusion is trivial. Now, suppose that m has some prime divisor p, greater than 1. Consider the $\mathbb{F}_p$-code C generated by the rows of U. In view of the equation $UU^T \equiv 0$ (mod p), the code C is self-orthogonal with respect to the dot product. However, C is reversible (since $p^j \equiv -1$ (mod exponent of G), for some integer j). By Proposition 4.8, the code C must be the zero code. Hence $U \equiv 0$ (mod p).

Let $U_1 = p^{-1}U$ and $m_1 = p^{-1}m$. Then $U_1U_1^T \equiv 0$ (mod m_1^2). By inductive hypothesis, $U_1 \equiv 0$ (mod m_1). Hence $U \equiv 0$ (mod m).□

Theorem 4.17. Let G *be an abelian group of order* w *and suppose that* A *is an integral* G-*matrix such that*

$$AA^T = xI + yJ$$

and

$$AJ = JA = zA$$

for some integers x,y,z. Suppose that for some integer m, we have $m^2 | x$ and m is semiprimitive (mod exponent of G). Then

$$A \equiv aJ \pmod{m}$$

where a is the solution to the congruence $wa \equiv z \pmod{m}$.

Proof. Set $U = A - aJ$, where a is an integer such that $wa \equiv z \pmod{m}$. Then

$$UU^T = xI + (wa^2 - 2za + y)J$$

$$\equiv (wa^2 - 2za + y)J \qquad \pmod{m^2}$$

since m^2 divides x. We claim that $wa^2 - 2za + y \equiv 0 \pmod{m^2}$.

To see this, multiply the equation $AA^T = xI + yJ$ by the matrix J to show that $z^2 = x + wy$ and hence $z^2 \equiv wy \pmod{m^2}$. Since m is semiprimitive (mod exponent of G) then certainly $(m,w) = 1$. Since $aw - z \equiv 0 \pmod{m}$, observe that

$$\begin{aligned} 0 &\equiv (aw-z)^2 && \pmod{m^2} \\ &\equiv wa^2 - 2az + w^{-1}z^2 && \pmod{m^2} \\ &\equiv wa^2 - 2az + w^{-1}(wy) && \pmod{m^2} \\ &\equiv wa^2 - 2az + y && \pmod{m^2} \end{aligned}$$

This shows that $UU^T \equiv 0 \pmod{m^2}$. The result now follows from Lemma 4.16.□

Theorem 4.17 applies directly to the contraction of the incidence matrix of an abelian difference set. In this section, we mention three nonexistence theorems which can be obtained in this way; there are certainly others.

Let us begin with a specific example. Suppose that there exists a (66,26,10)-difference set in a cyclic group G. Let H be a subgroup of order 2. Consider the contracted incidence matrix A_H. Its entries are 0,1,2 and it satisfies the equations

$$A_H A_H^T = 16\,I + 2.10\,J$$

and $$JA_H = A_H J = 26J.$$

nce $4^2|n$ and 4 is semiprimitive (mod exponent of G/H), we
nclude from Theorem 4.17 that

$$A_H \equiv 2J \pmod 4.$$

nce entries of A_H are 0,1, or 2, we must have $A_H=2J$. But
en A_H satisfies neither of the equations above! This
ntradiction proves that no such difference set exists. We
rase a general theorem.

Theorem 4.18. Suppose that there exists a (v,k,λ)-fference set in an abelian group G. Let H be a proper subоup of G, having order h. If for some integer m, we have

(1) m^2 divides n

(2) m is semiprimitive (mod exponent of G/H)

en $h \le m$.

Proof. Follow the model above. □

The table below gives examples of difference sets
cluded by this theorem.

,k,λ)	n	Group	m	h	exp(H)	$p^f \equiv -1$
6,11,2)	9	(8)(7)	3	2	28	$3^3 \equiv -1$ (mod 28)
0,13,4)	9	(2)(4)(5)	3	2	10	$3^2 \equiv -1$ (mod 10)
6,26,10)	16	(2)(3)(11)	4	2	33	$2^5 \equiv -1$ (mod 33)
04,29,4)	25	$(2)^2(3)(17)$	5	3	34	$5^8 \equiv -1$ (mod 34)
56,31,6)	25	$(2)^2(3)(13)$	5	3	26	$5^2 \equiv -1$ (mod 26)

Consider another example. Suppose that there exists
(40,13,4)-difference set in the group $G=\mathbb{Z}_2 \times \mathbb{Z}_4 \times \mathbb{Z}_5$.
t H be a subgroup of order 2 such that G/H has exponent 10.
have

$$A_H A_H^T \equiv 9I + 8J$$

and $\qquad A_H J = JA_H = 13J.$

We have $3^2|n$ and 3 is semiprimitive (mod 10). Thus,

$$A_H \equiv 2J \pmod{3}.$$

But, this is impossible. For, every entry of A_H must be at least 2 by virtue of the congruence and one entry must be at least 5 (since A_H could not be simply a multiple of J). But then the 20 x 20 matrix A_H has row sums at least 43, contrary to the fact that $A_H J=13J$. Thus no such difference set exists. We phrase a general theorem.

Theorem 4.19. Suppose that there exists a (v,k,λ)-difference set in an abelian group G. Let H be a proper subgroup of G, having order h and index w. Suppose that for some integer m,

(1) m^2 divides n

(2) m is semiprimitive (mod exponent of G/H).

Let a be the smallest nonnegative integer such that $aw \equiv k \pmod{m}$. Then

$$m \leq k-aw.$$

Proof. Follow the example above.□

The following table gives examples of difference sets excluded by the theorem.

(v,k,λ)	n	Group	h	w	exp(G/H)	m	a
(56,11,2)	9	(8)(7)	2	28	28	3	2
(40,13,4)	9	(2)(4)(5)	2	20	10	3	2
(154,18,2)	16	(2)(7)(11)	14	11	11	4	2
(78,22,6)	15	(2)(3)(13)	6	13	13	4	2

(66,26,10)	16	(2)(3)(11)	2	33	33	4	2
(204,29,4)	25	(4)(3)(17)	6	34	34	5	1
		$(2)^2(3)(17)$	6	34	34	5	1
(156,31,6)	25	$(2)^2(3)(13)$	3	52	26	5	3
(495,39,3)	36	(9)(5)(11)	5	99	99	2	1
		$(3)^2(5)(11)$	5	99	33	2	1
(1140,68,4)	64	(4)(3)(5)(19)	20	57	57	8	4
(806,70,6)	64	(2)(13)(31)	62	13	13	8	6

We consider one further example. Suppose that there exists a (154,18,2)-difference set in the abelian group $G = \mathbb{Z}_2 \times \mathbb{Z}_7 \times \mathbb{Z}_{11}$. Let H be a subgroup of order 14. Then A_H satisfies the equations

$$A_H A_H^T = 16\,I + 28\,J$$

$$\text{and} \quad A_J J = J A_H = 18J.$$

Since $4^2 | n$ and 4 is semiprimitive (mod 11), we have

$$A_J \equiv 2J \pmod 4.$$

Set $U = A_H - 2J$. Now

$$UU^T = (A_H - 2J)(A_H - 2J)^T = 16I$$

$$\text{and} \quad UJ = 16J.$$

By Lemma 4.17, we have $U \equiv 0 \pmod 4$. Set $E = \frac{1}{4}U$. Then

$$EE^T = I$$

$$\text{and} \quad EJ = 4J.$$

But this is impossible. For, by the first equation the sum of the squares of the (integral) entries in a row of E is 1, while the sum of these entries is 4 by the second equation. Thus, no such difference set exists. We phrase a general theorem.

Theorem 4.20. Suppose that there exists a (v,k,λ)-difference set in an abelian group G. Let H be a proper subgroup of G having order h and index w. Suppose that for some positive integer m,

(1) m^2 divides n

(2) m is semiprimitive (mod exponent of G/H).

Let a be any integer such that $aw \equiv k \pmod{m}$. Then

$$(k-aw)m \leq (a^2w-2ak+n+h\lambda).$$

Proof. Let $U=A_H-aJ$. Then

$$UU^T = nI + (h\lambda-2ak+a^2w)J$$

and $UJ = JU = (k-aw)J.$

As in the proof of Theorem 4.17, we have $UU^T \equiv 0 \pmod{m^2}$ and thus $U \equiv 0 \pmod{m}$. Set $E = \frac{1}{m}U$. Then E is an integral matrix such that

$$EE^T = \frac{1}{m^2}[nI + (h\lambda-2ak+a^2w)J]$$

and $EJ = JE = \frac{1}{m}[k-aw]J.$

For any row of E, the sum of the squares of the entries is as large as the sum of the entries. The theorem follows.□

The following table contains some examples of difference sets excluded by the theorem.

v,k,λ)	n	Group	h	w	exp(G/H)	m	a
56,11,2)	9	(8)(7)	2	28	28	3	-1
40,13,4)	9	(2)(4)(5)	2	20	10	3	-1
66,26,10)	16	(2)(3)(11)	2	33	33	4	-2

Remark. The inequality must hold for all values of such that $aw \equiv k \pmod m$. However, it is enough to consider ne single such value of a in the interval

$$\left(\frac{k}{w} - m\right) < a \leq \left(\frac{k}{w}\right).$$

To see this, move all terms of the inequality to the right-and side. We obtain a quadratic expression in a which is inimized at $a = \frac{k}{w} - \frac{m}{2}$.)

So far the major limitation of our techniques is the equirement that m be semiprimitive (mod exponent of G/H) in ur theorems. We can weaken this condition at the expense of lacing certain requirements on the structure of G. We do his in §§4.6-7. First, however, we digress to discuss a pecial topic.

4.5 DIFFERENCE SETS WITH MULTIPLIER -1

By this point it should be clear that -1 plays a pecial role in our nonexistence theorems. For, -1 refers to eversibility and reversibility entails severe restrictions.

Given the role of -1 in this theory, we should xpect that examples of difference sets with multiplier -1 hould be quite rare. This is indeed the case and we digress n this section to explore this special topic.

Only a few example of difference sets with -1 as a ultiplier are known:

(1) The difference sets of Examples 2 and 3 above admit -1 as a multiplier, as we observed in §4.1.

(2) Consider the difference sets in Example 4. By the construction, -1 will be a multiplier if and only if the multiplication table of K has the property that $xy=x^{-1}y^{-1}$, for all $x,y \in K$. That is, K must be an elementary abelian 2-group. This can occur whenever

$$q^d + \ldots + q + 2 = 2^u \qquad (*)$$

for some integer u. (We can always solve this for $q=2$, but this leads to the difference sets of example 2.) Only one solution is known for odd q, namely $(q,d)=(5,2)$. This yields a (4000,775,150)-difference set in the group $(\mathbb{Z}_5)^3 \times (\mathbb{Z}_2)^5$. It would be interesting to know whether (*) has any further solutions in prime powers q.

We can prove quite a lot about the structure of an abelian difference set admitting -1 as a multiplier.

Proposition 4.21. Suppose that D is a nontrivial symmetric (v,k,λ) design. An involution of D cannot fix exactly one point or exactly one block.

Proof. Let σ be an involution of D. It fixes an equal number of points and blocks. So, suppose that σ fixes just one point p and one block B. Let q be a nonfixed point not incident with B (which exists since D is nontrivial). The set of λ blocks through q and $\sigma(q)$ is permuted by σ; none of the blocks is fixed. Hence λ must be even. But, now let r be a point incident with B. The set of λ blocks through r and $\sigma(r)$ is permuted by σ; exactly one is fixed. Thus, λ must be odd. The contradiction proves the result.□

Corollary 4.22. Let $\mathcal{D}$ be a nontrivial (v,k,λ)-difference set in an abelian group G. If -1 is a multiplier of $\mathcal{D}$ then v is even and n is a square.

Proof. The multiplier -1 is an involution. In a group of odd order the map $g \longmapsto g^{-1}$ fixes exactly one element. Hence, G must have even order. By Schutzenberger's Theorem, n is a square.□

We can improve somewhat on Proposition 4.21.

Proposition 4.23. Suppose that D is a nontrivial mmetric (v,k,λ) design, with an involution σ fixing F points d blocks. If $F \neq 0$, then

$$F \geq \begin{cases} 1 + \frac{k}{\lambda} & \text{if } k \text{ and } \lambda \text{ are both even} \\ 1 + \frac{k-1}{\lambda} & \text{otherwise.} \end{cases}$$

Proof. Suppose that λ is even. Let B be a fixed ock. A pair of points of B fixed or interchanged by σ lie $\lambda-1$ further blocks, at least one of which is fixed. So ery point of B, with possibly one exception if k is odd, is cident with another fixed block. Now, each further fixed ock is incident with λ points of B. So, we must have

$$(F-1)\lambda \geq \begin{cases} k-1 & \text{if } k \text{ is odd} \\ k & \text{if } k \text{ is even.} \end{cases}$$

Suppose now that λ is odd. Let B be a fixed block. ch point outside B is incident with at least one fixed ock different from B. (For, if q is not fixed consider the blocks through q and $\sigma(q)$. If q is fixed, consider the λ ocks through q and any other fixed point of the design--of ich there must be at least one by Proposition 4.21.) Each xed block other than B is incident with $k-\lambda$ points outside Hence

$$(F-1)(k-\lambda) \geq v-k.$$

e result follows by Proposition 1.1 (3). □

Corollary 4.24. Let D be a nontrivial (v,k,λ)-fference set in an abelian group G. If -1 is a multiplier D, then G is not cyclic.

Proof. In a cyclic group, the map $g \mapsto g^{-1}$ fixes ther 1 or 2 elements. The previous proposition forbids is.□

Theorem 4.25. Suppose that $\mathcal{D}$ is a (v,k,λ)-differen set in an abelian group G which admits -1 as a multiplier. Let p be a prime dividing n. Then for some integer i, we have $p^{2i} \| n$, $p^i \| k$, $p^i \| \lambda$ and $p^{i+1} | v$.

Proof. Since n is a square, we may choose a positi integer i such that $p^{2i} \| n$. By the "moreover" portion of Theorem 4.14, we see that p must divide v. Say $p^a \| v$ and let P be the subgroup of G having order p^a. We have

$$A_P A_P^T = nI + p^a \lambda J.$$

Suppose that $a \le i$. Let p^e be the highest power of p dividing every entry of A_P. Certainly, $e \le a$. In fact, we claim that $e < a$. The following lemma proves this claim.

Lemma 4.26. Let $\mathcal{D}$ be a (v,k,λ)-difference set in a group G and let A be the incidence matrix of its developmen If H is a subgroup of order $h > 1$, then not every entry of A_H is a multiple of h.

Proof of lemma. By the construction of A_H, the entries are integers between 0 and h inclusive. If every entry of A_H is a multiple of h then every entry is either 0 or h. That is, $\mathcal{D}$ is a union of cosets of H. But then $g\mathcal{D} = \mathcal{D}$ for $g \in H$. This is impossible unless H is the identity subgroup.□

Let $B = p^{-e} A_P$. We have

$$BB^T = \frac{n}{p^{2e}} I + \frac{p^a \lambda}{p^{2e}} J.$$

In view of the equation $\lambda\lambda' = n(n-1)$, after possibly replacing $\mathcal{D}$ by its complement, we may assume that $p^i | \lambda$. (In a moment, we shall show that $p^i \| \lambda$ and thus $p^i \| \lambda'$. Accordingly, it does not matter whether we replace $\mathcal{D}$ by its complement here.) Every entry on the right-hand side is a multiple of p. Hence the rows of B generate a code over $\mathbb{F}_p$ which is self-orthogona with respect to the dot product. Since -1 is a G/P-multiplie code is reversible. By Proposition 4.8, we must have $B \equiv 0$ (moc this contradicts the choice of e as the largest integer h that p^e divides every entry of A_H. We have reached a tradiction by supposing that $a \leq i$. Hence $a \geq i+1$. Thus $^{1}|v$.

Now, $k \equiv \lambda \pmod{n}$. If $p^{i+1}|\lambda$, then it also divides Then $n=k^2-v\lambda$ must be a multiple of p^{2i+2}, contradicting choice of i. Hence $p^i \| \lambda$ and $p^i \| k$.□

Remarks. (1) The theorem shows that if $p^a \| v$ then +1. The (4000,775,150)-difference set shows that this nd can be attained.

(2) We can strengthen the conclusion of Theorem 5 as follows. If $p^{i+e} \| v$, then the exponent of the Sylow p-group of G is at most p^e. The proof requires techniques eloped later in this chapter. (See Problem 23.)

With only minor modifications the proof above can altered to prove a general result about difference sets which -1 is not necessarily a multiplier.

Theorem 4.27. Let $\mathcal{D}$ be a (v,k,λ)-difference set an abelian group G. Let p be a prime dividing n and v. K be a proper subgroup of G, containing the Sylow subgroup. Suppose there exists a numerical G/K-multiplier such that

$$t_{p^j} \equiv -1 \quad (\text{mod exponent of } G/K)$$

some integer j.

Then for some integer i, we have $p^{2i} \| n$, $p^i \| k$, $| \lambda$ and $p^i | v$. Moreover, if K is the Sylow p-subgroup of then in fact $p^{i+1} | v$.

Proof. The reader should adapt the proof of eorem 4.25.□

For example, Theorem 4.27 excludes the existence of (105,40,15)-difference set. A few examples of excluded fference sets are given in the table below.

(v,k,λ)	Group	|k|	exp (G/K)	p^j
(66,26,10)	(2)(3)(11)	2	33	$2^5 \equiv -1 \pmod{33}$
(120,35,10)	$(2)^3(3)(5)$	5	6	$5 \equiv -1 \pmod{6}$
(105,40,15)	(3)(5)(7)	5	21	$5^3 \equiv -1 \pmod{21}$

When −1 is a multiplier of an abelian difference set, it turns out that every other numerical automorphism must also be a multiplier. The proof relies on a basic property of cyclotomic fields. (See Appendix E.)

Theorem 4.28. Suppose that $\mathbb{D}$ is a (v,k,λ)-difference set in an abelian group G. If −1 is a multiplier of $\mathbb{D}$, then so is every integer t relatively prime to v. Moreover, if $\mathbb{D}$ happens to be fixed by the multiplier −1 then $\mathbb{D}$ is fixed by every numerical automorphism.

Proof. Suppose that −1 is a multiplier. Since this multiplier fixes at least one point, it must fix at least one block. So, after possibly replacing the difference set by a translate, we may assume that −1 fixes the difference set $\mathbb{D}$. Let us use the notation of the group ring $\mathbb{Z}G$. From Proposition 4.3, we have

$$\underline{\mathbb{D}}^2 = \underline{\mathbb{D}}\,\underline{\mathbb{D}}^{(-1)} = n\underline{1}_G + \lambda \underline{G}.$$

Let χ be any complex-valued character of G. Then

$$\chi(\underline{\mathbb{D}})^2 = n + \lambda\chi(\underline{G}) = \begin{cases} n + \lambda v = k^2 & \text{if } \chi \text{ is principal} \\ n & \text{if } \chi \text{ is nonprincipal} \end{cases}$$

by Proposition D.6. By Corollary 4.22, the integer n is a square. Hence, for every complex-valued character χ, the number $\chi(\underline{\mathbb{D}})$ is an integer.

Now, the characters χ take values in the field $\mathbb{L} = Q(\zeta)$ where ζ is a primitive v-th root of unity. Let σ_t be the automorphism of $\mathbb{L}$ carrying ζ to ζ^t, where t is an nteger relatively prime to v. Since σ_t fixes the rational ubfield of L, then

$$\chi(\underline{D}) = \sigma_t(\chi(\underline{D}))$$
$$= \chi(\underline{D}^{(t)}),$$

or all characters χ. Thus all characters take the same alue on $\underline{D}$ and $\underline{D}^{(t)}$. By §D.4, we must have $\underline{D}=\underline{D}^{(t)}$. Hence t s a multiplier.□

Remark. A similar result holds when -1 is a G/H-ultiplier, but it is slightly more complicated:

(1) Provided that n is a square, the proof above ıffices to show that every numerical automorphism of G/H a G/H-multiplier.

(2) Suppose that n is not a square. Then $n^*=p$ for me prime $p\equiv 1 \pmod 4$ and G/H is a p-group, by Corollary 4.15. y $|G/H|=p^f$. Let us mimic the proof above. Let $L=Q(\zeta)$, ere ζ is a primitive p^f-th root of unity. The Galois oup Gal(L,Q) is isomorphic to the group of units (mod p^f), ich is a cyclic group. (See Problem 8 of Chapter 5.)

Now, for each character χ, the number $\chi(\underline{D})$ is the uare root of an integer--that is, $\chi(\underline{D})$ lies in a quadratic ofield of L. By Galois theory, a quadratic subfield of L the fixed field of a subgroup of index 2 in Gal(L,Q), which necessarily the subgroup of squares. Hence σ_t fixes $\chi(\underline{D})$ enever t is a nonzero square (mod p). Thus the numerical H multipliers are precisely the nonzero quadratic residues od p).

6 CYCLIC GROUPS ARE SPECIAL

We return now to our main task of proving non-ıstence theorems for abelian difference sets. Virtually ery nonexistence theorem so far has rested upon an applica-on of Proposition 4.8. That fundamental result depended turn on the very simple structure of the abelian group g $F_p[G]$, when p does not divide |G|. Clearly it would be desirable to eliminate, or at least weaken, the requirement on p. Unfortunately, the structure of $F_p[G]$ is quite unruly in general when p divides $|G|$. There is, however, an important exception. When the Sylow p-subgroup of G is cyclic, it is not too difficult to determine the ideal structure of the group ring. This topic is discussed in §D.6 of Appendix D, which the reader should skim now.

In the sequel, fix the following notation. Let p be a prime. Suppose that G is an abelian group with a non-trivial Sylow p-subgroup P, which is cyclic. Say $G = P \times Q$. Let P' be the unique subgroup of P having order p. Let μ be the canonical projection from $G=P\times Q$ to $G/P' = (P/P')\times Q$. We may extend μ to a map

$$\mu : F_p[G] \longrightarrow F_p[G/P']$$

by linearity. We have the following result.

<u>Proposition 4.29</u>. <u>Suppose that</u> p <u>is semiprimitive</u> (mod exponent of Q). <u>Then every</u> G-<u>code self-orthogonal (with respect to the dot product) lies in the kernel of</u> μ.

<u>Proof</u>. Let $m = |P|$. According to §D.6 of Appendi[x] every G-code C in $F_p[G] = F_p[P] \otimes F_p[Q]$ can be decomposed as

$$C = \bigoplus_{i=1}^{s} (V_{d_i} \otimes U_i)$$

where the U_i are the indecomposable $F_p[Q]$-modules and V_d denotes the unique ideal of $F_p[P]$ having dimension d. We also noted that

$$C^* = \bigoplus_{i=1}^{s} (V_{d_i} \otimes U_i^*).$$

From the proof of Proposition 3.12, we have $C^* \simeq V/C^{\perp}$, where we write V for $F_p[G]$. Since

$$V = \bigoplus_{i=1}^{s} (V_m \otimes U_i^*)$$

en $C^{\perp} = \bigoplus_{i=1}^{s} (V_{m-d_i} \otimes U_i^{*})$.

Now suppose that p is semiprimitive (mod exponent Q) and that C is a self-orthogonal code. The condition on implies that each of the U_i is self-contragredient. Thus,

$$C^{\perp} = \bigoplus_{i=1}^{s} (V_{m-d_i} \otimes U_i).$$

e condition that $C \subseteq C^{\perp}$ now implies that $V_{d_i} \subseteq V_{m-d_i}$ for ,...,s. In other words, $d_i \le \frac{1}{2}m$.

In Appendix D, we noted that the kernel of the pro- ction $\pi: F_p[P] \longrightarrow F_p[P/P']$ is $V_{(\frac{p-1}{p})m}$. Since $\frac{1}{2} \le \frac{p-1}{p}$, then i lies in the kernel of π, for i=1,...,s. Hence

$$\mu(C) = \bigoplus_{i=1}^{s} (\pi(V_{d_i}) \otimes U_i) = 0.$$

is completes the proof.□

It is interesting to compare Propositions 4.8 and 29. The former asserts that (under appropriate hypotheses) self-orthogonal code is necessarily the zero code. The tter is slightly weaker, but in the same spirit. It asserts at (under appropriate hypotheses) a self-orthogonal code comes the zero code when contracted by a subgroup of order p.

We now prove the main theorem of this section. In eparation, we make the following definition. Call a prime *self-conjugate* (mod w) if p is semiprimitive (mod w_p), ere w_p is the largest divisor of w relatively prime to p. ll a composite number m *self-conjugate* (mod w) if every ime factor of m is self-conjugate (mod w).

Theorem 4.30. *Suppose that there exists a* (v,k,λ)-*fference set* D *in an abelian group* G. *If* p *is a prime such at*

(1) p *divides* v *and* n, *and*

(2) p *is self-conjugate* (mod exponent of G) *then the Sylow* p-*subgroup of* G *is not cyclic*.

Proof. Suppose that $\mathcal{D}$ is a (v,k,λ)-difference set in an abelian group G satisfying the hypotheses of the theorem and suppose that the Sylow p-subgroup of G is cyclic. Let A be the incidence matrix of dev $(\mathcal{D})$. We have

$$AA^T = nI + \lambda J.$$

In view of the equation $n=k^2-v\lambda$, we see that p divides k and hence also λ. So, the $\mathbb{F}_p$-code C spanned by the rows of A is self-orthogonal with respect to the dot product.

We can apply Proposition 4.29. The code $\mu(C)$, obtained by contracting the subgroup P' of order p, must be the zero code. But $\mu(C)$ is simply the $\mathbb{F}_p$-span of the rows of $A_{P'}$. Hence,

$$A_{P'} \equiv 0 \pmod p)$$

Therefore every entry of $A_{P'}$ is a multiple of p. But, this contradicts Lemma 4.26. We conclude that the Sylow p-subgroup of G cannot be cyclic.□

Theorem 4.30 rules out many putative difference sets. For example, there does not exist a cyclic (16,6,2)-difference set since 2 is self-conjugate (mod 16). The table below gives further examples ruled out by the theorem.

(v,k,λ)	n	Group	$p^f \equiv -1$	(mod w)
(16,6,2)	4	(16)	$2 \equiv -1$	(mod 1)
(45,12,3)	9	(9)(5)	$3^2 \equiv -1$	(mod 5)
(36,15,6)	9	(4)(9)	$2^3 \equiv -1$	(mod 9)
		$(2)^2(9)$	$2^3 \equiv -1$	(mod 9)
(96,20,4)	16	(32)(3)	$2^1 \equiv -1$	(mod 3)
(66,26,10)	16	(2)(3)(11)	$2^5 \equiv -1$	(mod 33)
(64,28,12)	16	(64)	$2^1 \equiv -1$	(mod 1)
(175,30,5)	25	(25)(7)	$5^3 \equiv -1$	(mod 7)

(v,k,λ)	n	Group	$p^f \equiv -1$	(mod w)
(120,35,10)	25	$(2)^3(3)(5)$	$5^1 \equiv -1$	(mod 6)
(704,38,2)	36	(64)(11)	$2^5 \equiv -1$	(mod 11)
(105,40,15)	25	(3)(5)(7)	$5^3 \equiv -1$	(mod 21)
(288,42,6)	36	(32)(9)	$2^3 \equiv -1$	(mod 8)
		$(32)(3)^2$	$2^1 \equiv -1$	(mod 3)
(111,45,18)	27	(3)(37)	$3^9 \equiv -1$	(mod 37)
(100,45,20)	25	$(2)^2(25)$	$5^1 \equiv -1$	(mod 2)
(208,46,10)	36	(16)(13)	$2^6 \equiv -1$	(mod 13)
(189,48,12)	36	(27)(7)	$3^3 \equiv -1$	(mod 7)
(176,50,14)	36	(16)(11)	$2^5 \equiv -1$	(mod 11)
(171,51,15)	36	(9)(19)	$3^9 \equiv -1$	(mod 9)
(160,54,18)	36	(32)(5)	$2^2 \equiv -1$	(mod 5)
(153,57,21)	36	(9)(17)	$3^8 \equiv -1$	(mod 17)
(280,63,14)	49	$(2)^3(5)(7)$	$7^2 \equiv -1$	(mod 10)
(2146,66,2)	64	(2)(29)(37)	$2^{126} \equiv -1$	(mod 1073)
(144,66,30)	36	(16)(9)	$2^3 \equiv -1$	(mod 8)
		$(16)(3)^2$	$2^1 \equiv -1$	(mod 3)
		$(9)(4)^2$	$3^1 \equiv -1$	(mod 4)
		$(9)(4)(2)^2$	$3^1 \equiv -1$	(mod 4)
		$(9)(2)^4$	$3^1 \equiv -1$	(mod 2)
(783,69,6)	63	(27)(29)	$3^{14} \equiv -1$	(mod 29)
(640,72,8)	64	(128)(5)	$2^2 \equiv -1$	(mod 5)

The condition that p is self-conjugate (mod exponent of G) is clearly crucial to the proof of Theorem 4.30. Yet, it is not clear to me that the condition is necessary at all. I conjecture, on the basis of many examples, that Theorem 4.30 remains true without this condition. We shall see evidence for this conjecture in Chapter 6.

While I do not know how to prove this conjecture, I can show, in the case that p=2, how to substitute a modest divisibility condition for condition (2) in Theorem 4.30.

Theorem 4.31. *Suppose that there exists a* (v,k,λ)-*difference set in an abelian group G. If*

(1) v *and* n *are even, and*

(2) $2^{2i} \| n$ *and* $2^i \| k$ *for some positive integer* i,

then the Sylow 2-subgroup of G is not cyclic.

Proof. We begin with an observation about divisibility. If $2^{2i} \| n$ and $2^i \| k$ then $2^i \| \lambda$. By virtue of the equation $n=k^2-v\lambda$, then also $2^i | v$.

Suppose now that $\mathcal{D}$ is a (v,k,λ)-difference set in an abelian group G satisfying the hypotheses of the theorem. Write G=PxQ, where P is the Sylow 2-subgroup of G. Suppose that P is cyclic. Let A be the incidence matrix of dev $(\mathcal{D})$. The contracted matrix A_Q satisfies the equations

$$A_Q A_Q^T = nI + q\lambda J$$

and $$JA_Q = A_Q J = kJ.$$

where $q=|Q|$. The $\mathbb{F}_2$-code spanned by the rows of A_Q is self-orthogonal with respect to the dot product, and has a cyclic 2-group acting on it. Let A_1 be the matrix obtained by contracting A_Q by a subgroup of order 2. Then A_1 satisfies the equations,

$$A_1 A_1^T = nI + 2q\lambda J$$

and $$JA_1 = A_1 J = kJ$$

and also $A_1 \equiv 0 \pmod 2$, by Proposition 4.29. Set $B_1 = \frac{1}{2}A_1$. Then B_1 is an integral matrix such that

$$B_1 B_1^T = \frac{n}{4} I + \frac{q\lambda}{2} J$$

and $$JB_1 = B_1 J = \frac{k}{2} J .$$

Repeat the process. Let A_2 be the matrix obtained by contracting B_1 by a subgroup of order 2. Then $A_2 \equiv 0 \pmod 2$. Set $B_2 = \frac{1}{2}A_2$. Continue in this way, obtaining an integral matrix B_i such that

$$B_i B_i^T = \frac{n}{2^{2i}} I + \frac{q\lambda}{2^i} J$$

and $$JB_i = B_i J = \frac{k}{2^i} J .$$

Now, $2^{-i}k$, $2^{-2i}n$ and $2^{-i}q\lambda$ are all odd integers. Thus, the second equation says that the sum of the entries of B_i is odd and the first says that the sum of the squares of the entries is even. This is impossible. Hence P is not cyclic. □

Theorem 4.31 rules out many potential difference sets. For example, there are no cyclic $(4N^2, 2N \pm N, N^2 \pm N)$-difference sets whenever N is even. (The parameters are those associated with Hadamard matrices with constant row and column sums.)

We close this section with another attempt at weakening condition (2) of Theorem 4.30. It is best to start with an example.

Consider a (288,42,6)-difference set $\mathcal{D}$ in an abelian group G. We have $v=288=2^5 3^2$. The Sylow 2-subgroup of G cannot be cyclic by Theorem 4.31. What about the Sylow 3-subgroup? Since 3 is not necessarily self-conjugate (mod exponent of G), Theorem 4.30 does not directly apply. So, suppose that the Sylow 3-subgroup is cyclic.

Let H be a subgroup of G having order 4 such that G/H has exponent dividing 36 (which must exist since the Sylow 2-subgroup of G is not cyclic). If A is the incidence matrix of dev (D) then the contraction A_H satisfies the equations

$$A_H A_H^T = 36I + 24J$$

$$\text{and} \quad A_H J = J A_H = 42J.$$

Since 3 is self-conjugate (mod 36), Proposition 4.29 applies. If A_1 is the matrix obtained by contracting A_H by a subgroup of order 3, then $A_1 \equiv 0 \pmod 3$. Set $B_1 = \frac{1}{3}A_1$. Then B satisfies the equations

$$BB^T = 4I + 8J$$

$$\text{and} \quad BJ = JB = 14J.$$

We now have a contradiction. For, the sum of the entries of each row of B is 14 while the sum of the squares of the entries is 12. Hence the Sylow 3-subgroup of G cannot be cyclic. We have the following result.

<u>Theorem 4.32. Suppose that there exists a</u> (v,k,λ)-<u>difference set in an abelian group G. Let H be a subgroup of order h and index w. Suppose that for some positive integer m,</u>

(1) m^2 <u>divides</u> n

(2) m <u>divides</u> k, λ <u>and</u> w

(3) m <u>is self conjugate</u> (mod exponent of G/H) <u>and</u>

(4) <u>if</u> p <u>is a prime dividing</u> m <u>and</u> w, <u>then the Sylow</u> p-<u>subgroup of</u> G/H <u>is cyclic</u>.

<u>Then</u>

$$h \geq \frac{1}{m} + \left(\frac{k}{\lambda}\right)\left(\frac{m-1}{m}\right).$$

<u>Proof</u>. Let D be a difference set satisfying the hypotheses of the theorem and let A be the incidence matrix of dev (D). Consider the matrix A_H. We successively contract

A_H by groups of prime order until we have contracted it by the (unique) subgroup of order m; each time we contract by a group of prime order p the resulting matrix is a multiple of p. Thus, just as in the proof of Theorem 4.31, we obtain an integral matrix E such that

$$EE^T = \frac{n}{m^2} I + \frac{h\lambda}{m} J$$

$$\text{and} \quad EJ = JE = \frac{k}{m} J.$$

Since the sum of the squares of the entries of a row of E must be no less than the sum of the entries, we have $mk \leq n + mh\lambda$. Or, equivalently,

$$h \geq \frac{1}{m} + \left(\frac{k}{\lambda}\right)\left(\frac{m-1}{m}\right).$$

This completes the proof.□

Notice that Theorem 4.32 includes Theorem 4.30 as a special case (when h=1). Theorem 4.32 can be used also to rule out further difference sets. In the example above, we ruled out an abelian (288,42,6)-difference set in a group with a cyclic Sylow 3-subgroup. (Here h=4 and m=3.) The following table lists further examples excluded by the theorem.

(v,k,λ)	n	Group	h	exp G/H	m
(96,20,4)	16	(2)(16)(3)	2	48	4
(120,35,10)	25	(2)(4)(3)(5)	2	60	5
(704,38,2)	36	(2)(32)(11)	8	88	2
		(4)(16)(11)	8	88	2
		$(2)^2$(16)(11)	8	88	2
		$(8)^2$(11)	8	88	2
		(2)(4)(8)(11)	8	88	2

(v,k,λ)	n	Group	h	exp G/H	m
		$(2)^3(8)(11)$	8	88	2
(288,42,6)	36	(2)(16)(9)	4	36	3
		(4)(8)(9)	4	36	3
		$(2)^2(8)(9)$	4	18	3
		$(2)(4)^2(9)$	4	36	3
		$(2)^5(9)$	4	18	3
		$(2)(16)(3)^2$	2	48	2
		$(2)(4)^2(3)^2$	2	12	3
		$(2)^3(4)(3)^2$	3	12	3
		$(2)^5(3)^2$	3	6	3
(208,46,10)	36	(2)(8)(13)	2	104	2
(176,50,14)	36	(2)(8)(11)	2	88	2

§4.7 MORE ON CYCLIC GROUPS

Continuing our exploration of cyclic subgroups, we prove the following result.

Theorem 4.33. Suppose that there exists a (v,k,λ)-difference set in an abelian group G. Let H be a subgroup of G of order h and index w. Suppose that m is an integer such that

(1) m^2 divides n,

(2) $(m,w)\neq 1$,

(3) m is self-conjugate (mod exponent of G/H),

(4) if p is a prime dividing m and w, then the Sylow p-subgroup of G/H is cyclic.

Then $m \leq 2^{r-1}h,$

where r is the number of distinct prime factors of (m,w).

Before turning to the proof of Theorem 4.33, we explore some of its consequences. Theorem 4.33 includes Theorem 4.30 as a special case (take h=1), but can be used to exclude further difference sets. For example it rules out a (64,28,12)-difference set in $\mathbb{Z}_{32} \times \mathbb{Z}_2$. The following table gives further examples of difference sets excluded by Theorem 4.33.

(v, k, λ)	n	Group	h	exp G/H	m
(96,20,4)	16	(2)(16)(3)	2	48	4
(78,22,6)	16	(2)(3)(13)	3	26	4
(64,28,12)	16	(2)(32)	2	32	4
(120,35,10)	25	(8)(3)(5)	4	30	5
		(2)(4)(3)(5)	2	60	5
(100,45,20)	25	(4)(25)	2	50	5
(280,63,14)	49	(8)(5)(7)	5	56	7
		(2)(4)(5)(7)	5	28	7
(144,66,30)	36	(9)(8)(2)	2	36	3
(231,70,21)	49	(3)(7)(11)	3	77	7

Theorem 4.33 is particularly useful in the case of $(4N^2,\ 2N^2 \pm N,\ N^2 \pm N)$-difference sets, those associated with Hadamard matrices having constant row and column sums. For groups G and primes p, define the function σ_p as follows: $\sigma_p(G)=a$ if a maximal cyclic p-subgroup of G has order p^a.

Using this notation, we have the following corollary to Theorem 4.33.

Corollary 4.34. Suppose that there exists a $(4N^2, 2N^2 \pm N, N^2 \pm N)$-difference set in an abelian group G.

(1) If $N=2^a$ then $\sigma_2(G) \leq a+2$.

(2) If $N=p^a$ for an odd prime p, then $\sigma_p(G) \leq a$.

Proof. (1) Let $\sigma_2(G)=b$. Let H be a subgroup of order 2^{2a+2-b} such that $\sigma_2(G/H)=b$. Applying Theorem 4.33 with $m=2^a$, we have $2^a \leq 2^{2a+2-b}$. Hence $b \leq a+2$.

(2) Suppose that $p>2$. Let $\sigma_p(G)=b$. We can find a subgroup H of order $2p^{2a-b}$ such that $\sigma_p(G/H)=b$. Since $|G/H|=2p^b$ and p is semiconjugate (mod $2p^b$), we can apply Theorem 4.33 with $m=p^a$. We have $p^a \leq 2p^{2a-b}$. Thus, $p^b \leq 2p^a$ and $b \leq a$. □

The case of $(2^{2m+2}, 2^{2m+1} \pm 2^m, 2^{2m} \pm 2^m)$-difference sets, those with $N=2^m$ in the corollary above, is particularly interesting. By the corollary, it is necessary that G have exponent at most 2^{m+2}. By Example 4½ of §4.1, there is a difference set attaining this bound in $(\mathbb{Z}_{2^{m+2}}) \times (\mathbb{Z}_2)^m$. A wide gap, however, separates the necessary condition in Corollary 4.34 and the sufficient condition in Example 4½. It would be interesting to know the answer to the following question:

Question: Does there exist a $(2^{2m+2}, 2^{2m+1} \pm 2^m, 2^{2m} \pm 2^m)$-difference set in an abelian group G of exponent $\leq 2^{m+2}$ and rank $\leq m$?

We now proceed to prove Theorem 4.32, which is due to Turyn [132]. Rather than use our approach based on codes and modules we shall use Turyn's original proof, which depends on algebraic number theory and characters. For this particular result the latter method is more straightforward and, in any case, it is desirable to present one example using the sort of character-theoretic argument found frequently in the literature.

We shall require three lemmas which rely on the prime factorization of ideals in cyclotomic fields. (This topic is discussed in Appendix E.) Consider the group ring

P, where P is a cyclic group of order p^a, with p a prime.
et χ be a complex-valued character of P carrying a fixed
enerator g to a fixed p^a-th root of unity, ζ. Then χ induces
ne map

$$\chi: \mathbb{Z}P \longrightarrow \mathbb{Z}[\zeta]$$

hose kernel is the ideal $(1+g^{p^{a-1}}+\dots+g^{p^{a-1}(p-1)})$, since
$(x)=1+x^{p^{a-1}}+\dots+x^{\phi(p^a)}$ is the monic minimal polynomial of ζ.
et us be more explicit about the map. If

$$s = \sum_{i=0}^{p^a-1} a_i g^i \in \mathbb{Z}P$$

hen

$$\chi(s) = \sum_{i=0}^{\phi(p^a)-1} b_i \zeta^i$$

ith $b_i=(a_i-a_{t(i)})$, where $t(i)$ is the integer such that
$(p^a)\le t(i)\le p^a-1$ and $t(i)\equiv i \pmod{p^{a-1}}$. In particular, notice
he following facts.

(1) If $a_0\equiv a_1\equiv\dots\equiv a_{p^a-1} \pmod{m}$, for some integer m,
hen m divides all the b_i and hence $\chi(s)$.

(2) If for some integer M, we have $0\le a_i\le M$ for
$=0,1,\dots,p^a-1$, then $-M\le b_i\le M$. And if $-M\le a_i\le M$, then $-2M\le b_i\le 2M$.

We now prove three lemmas.

Lemma 4.35. Let $s \in \mathbb{Z}P$. If $\chi(ss^{(-1)})\equiv 0 \pmod{p^{2i}}$,
hen $\chi(s)\equiv 0 \pmod{p^i}$.

Proof. We have $\chi(ss^{(-1)})=\chi(s)\chi(s^{(-1)})=\chi(s)\overline{\chi}(s)$
here the bar denotes complex conjugation. Suppose that
$^{2i}|\chi(s)\overline{\chi}(s)$. Now, in $Q(\zeta)$, the ideal (p) is a power of a
rime ideal π, which is necessarily fixed by complex conjuga-
ion. Say $p=\pi^e$. Then π^{2ie} divides $(\chi(s)\overline{\chi}(s))$. Since $\pi=\overline{\pi}$,
he same power of π divides $(\chi(s))$ and $(\overline{\chi}(s))$ exactly. Hence
e must have $\pi^{ie}|(\chi(s))$ and thus $\chi(s)\equiv 0 \pmod{p^i}$.□

Next consider the group ring $\mathbb{Z}G$, where $G= P \times Q$,
ith P as above and Q an abelian group w relatively prime to

p. Let $q_1, \ldots, q_w$ be the elements of Q. We can express every element $s \in \mathbb{Z}G$ in the form $s = \sum_{j=1}^{w} s_j q_j$, where $s_j \in \mathbb{Z}P$. Moreover, extend χ to a character on all of G by letting χ act as the principal character on Q. Similarly, if η is a character of Q, we will extend η to be a character of G by letting it act as the princiapl character on P.

<u>Lemma 4.36</u>. Let $s \in \mathbb{Z}G$ and write $s = \sum_{j=1}^{w} s_j q_j$, with $s_j \in \mathbb{Z}P$. Suppose that $(\chi\eta)(ss^{(-1)}) \equiv 0 \pmod{p^{2i}}$ for all characters η of Q. Also, suppose that p is self-conjugate (mod exponent of G). Then $\chi(s_j) \equiv 0 \pmod{p^i}$ for $j=1,\ldots,w$.

<u>Proof</u>. We have $\chi\eta(s)\overline{\chi\eta}(s) \equiv 0 \pmod{p^{2i}}$, where the characters take values in the field of v-th roots of unity, with v=exponent of G. By Theorem E.2, the condition that p is self-conjugate (mod exponent of G) is equivalent to the fact that the ideal (p) factors as a product of prime ideals all fixed by complex conjugation. Arguing as in the lemma above, we have $(\chi\eta)(s) \equiv 0 \pmod{p^i}$ for all characters η of Q. Thus,

$$\sum_{j=1}^{w} \eta(q_j)\chi(s_j) \equiv 0 \pmod{p^i}$$

for all characters η of Q. Express this as a matrix equation,

$$A \begin{pmatrix} \chi(s_1) \\ \cdot \\ \cdot \\ \cdot \\ \chi(s_w) \end{pmatrix} \equiv \begin{pmatrix} 0 \\ \cdot \\ \cdot \\ \cdot \\ 0 \end{pmatrix} \pmod{p^i}$$

where A is the character table of Q. Since A is invertible (mod p^i) then $\chi(s_j) \equiv 0 \pmod{p^i}$ for $j=1,\ldots,w$.□

<u>Lemma 4.37</u>. <u>Let $G = P_1 \times \ldots \times P_r \times Q$ be an abelian group where P_i is a cyclic p_i-group and where $(p_1 \ldots p_r, |Q|) = 1$.</u> <u>Let m be an integer whose prime factors are exactly $p_1, \ldots, p_r$.</u>

Let s be an element of $\mathbb{Z}G$ whose coefficients all e in the interval $[0,M]$ and suppose that $\chi(ss^{(-1)})\equiv 0$ od m^2) for all nonprincipal characters of G. Suppose also at $\chi_0(s)\neq 0$, where χ_0 is the principal character of G. If is self-conjugate (mod exponent of G) then $m<2^{r-1}M$.

Proof. Let χ_i be a character of P_i, carrying a nerator g_i of P_i into a given primitive $|P_i|$-th root of ity, ρ_i. Consider the map

$$\theta = \chi_1 x\ldots x\chi_r \times id_Q: \mathbb{Z}G \longrightarrow \mathbb{Z}[\zeta_1]x\ldots x\,\mathbb{Z}[\zeta_r] \times \mathbb{Z}Q$$

By the previous lemma we have $\theta(s)\equiv 0$ (mod m). However, by our initial remarks the coefficients of $\theta(s)$ lie in e interval $[-2^{r-1}M,\ 2^{r-1}M]$. Since $\chi_0(s)\neq 0$, not all of these efficients are zero. Since they are all multiples of m, en $m\leq 2^{r-1}M$.□

Proof of Theorem 4.33. Let A be the incidence trix of the development of the difference set. Then A_H tisfies the equations

$$A_H A_H^T = nI + h\lambda J.$$

$$\text{and}\quad A_H J = JA_H = kJ.$$

nsider a row of A_H as an element of $\mathbb{Z}[G/H]$. The coefficients of s lie in the interval $[0,h]$ and we have

$$s\,s^{(-1)} = n\underline{1} + h\lambda \underline{G/H}.$$

r all nonprincipal characters χ of G/H, we have $\chi(\underline{G/H})=0$ hy?) and hence $\chi(ss^{(-1)})=n$.

Write m_1m_2 where every prime factor of m_1 divides and no prime factor of m_2 divides w. By Theorem 4.17, the efficients of s are all congruent (mod m_2) and hence $\chi(s)\equiv 0$ od m_2), as we noticed when we defined the map χ above. t $s' = \frac{1}{m_2}s$. The coefficients of s' lie in the interval ,h/m_2]. Applying Lemma 4.37 directly to s', we conclude that

$$m_1 \leq 2^{r-1}(h/m_2),$$

where r is the number of distinct primes dividing m_1. Hence $m \leq 2^{r-1}h$. This completes the proof.□

§4.8 FURTHER RESULTS

To conclude our exploration of nonexistence theorems for abelian difference sets, we mention without proof four further results.

Theorem 4.38. Suppose that there exists a (v,k,λ)-difference set in an abelian group G. Let q be a prime congruent to 3 (mod 4) and suppose that G has a cyclic subgroup of order q^ℓ for some integer $\ell>0$. Suppose that every prime divisor p of n satisfies one of the following conditions:

(1) $\mathrm{ord}_q(p)$ is even

(2) $\mathrm{ord}_{q^\ell}(p) = \frac{1}{2}\phi(q^\ell)$, (the Euler ϕ-function)

(3) $p=q$.

Then there must exist integers x and y such that

$$4n=x^2+qy^2,\ 0\leq x,\ 0\leq y\leq q^{-\ell}v \text{ and } x+y\leq 2q^{-\ell}v.$$

Theorem 4.39. Suppose that there exists a (v,k,λ)-difference set in an abelian group G. Let q and r be primes with $q\equiv 3 \pmod 4$. Suppose that G has cyclic subgroups of order q^ℓ and r^m and suppose that $(\phi(q^\ell), \phi(r^m))=2$. Furthermore, suppose that every prime p dividing n satisfies one of the following conditions:

(1) $\mathrm{ord}_q(p)$ is even and $\mathrm{ord}_r(p)\equiv 2 \pmod 4$

(2) $\mathrm{ord}_{q^\ell}(p) = \frac{1}{2}\phi(q^\ell)$ and $\mathrm{ord}_{r^m}(p) = \phi(r^m)$

(3) $p=q$ and $\mathrm{ord}_{r^m}(p) = \phi(r^m)$.

Then there must exist integers x and y such that

$$4n=x^2+qy^2,\ 0\leq x,\ 0\leq y\leq 2q^{-\ell}r^{-m}v \text{ and } x+y\leq 4q^{-\ell}r^{-m}v.$$

Theorem 4.40. Suppose that there exists a (v,k,λ)-
:ference set in an abelian group G. Let q and r be primes
h $q\equiv 3 \pmod 4$, $r\equiv 1 \pmod 4$ and q a nonsquare (mod r).
pose that G has cyclic subgroups of order q^{ℓ} and r^m and
pose that $(\phi(q^{\ell}), \phi(r^m))=2$. Furthermore, suppose that
ry prime p dividing n satisfies one of the following
ditions:

(1) $ord_q(p)$ is even and $ord_r(p)\equiv 2 \pmod 4$

(2) $ord_{q^\ell}(p) = \phi(q^{\ell})$ and $ord_{r^m}(p) = \phi(r^m)$

n there must exist integers x and y such that

$$4n=x^2+qry^2,\quad 0\le x,\quad 0\le y\le 2q^{-\ell}r^{-m}v \text{ and } x+y\le 4q^{-\ell}r^{-m}v.$$

These results, due to Yamamoto [143], are proven
ng character-theoretic counting. The various number-
oretic assumptions turn out to be conditions forcing all
evant character sums to lie in an imaginery quadratic
ld (and thus to be relatively tractable). Yamamoto only
tes the results for cyclic difference sets but his proofs
fice for the more general versions stated above.

We mention a few applications of these results.
orem 4.38 rules out a (239,35,5)-difference set. (Notice
t since v=q=239, there are only five pairs (x,y) which
n satisfy the last three inequalities!) Theorem 4.39
ludes a cyclic (306,61,12)-difference set, taking $q^{\ell}=9$ and
=17. Theorem 4.40 excludes a (286,96,32)-difference set.
table below lists examples excluded.

$v,k,\lambda)$	n	Group	q^{ℓ}	r^m	Theorem
27,13,6)	7	(27)	3^3	—	4.38
115,19,3)	16	(5)(23)	23^1	5^1	4.39
239,35,5)	30	(239)	239^1	—	4.38

(v,k,λ)	n	Group	q^{ℓ}	r^m	Theorem
(99,49,24)	25	(9)(11)	11^1	3^2	4.39
(306,61,12)	49	(2)(9)(17)	3^2	17^1	4.39
(286,96,32)	64	(2)(11)(13)	11^1	13^1	4.40

We close this chapter with results due to Mann [92]. Whereas Yamamoto's methods concentrated on cyclic groups, Mann's go to the opposite extreme, by working with elementary abelian groups.

Theorem 4.41. Suppose that there exists a (v,k,λ)-difference set in an abelian group G. Let H be a subgroup of G having order h and such that G/H is an elementary abelia group of order p^m with p an odd prime.

If the quadratic residues (mod p) are G/H-multiplie then

$$x^2 + py^2 = 4n \qquad (*)$$

has an integral solution. Moreover, let $(x_1,y_1),(x_2 y_2),\ldots,(x_\ell,y_\ell)$ be the solutions of (*) which also satisfy the additional conditions

(1) $2k \equiv x \pmod p$,

(2) $k + \frac{1}{2}(p-1)x \geq 0$,

(3) $k \geq \frac{1}{2}(x+|y|p)$.

Then $\ell>0$ and the following equations

$$k + \tfrac{1}{2}(p-1)(x_1 z_1 + \ldots + x_\ell z_\ell) = jp^m$$

$$z_1 + \ldots + z_\ell = \left(\frac{p^m-1}{p-1}\right)$$

have a solution in nonnegative integers $z_1,\ldots,z_\ell,j$ with $j \leq h$.

The proof is also by character-theoretic counting (Mann only states the result in the case that H is the identi subgroup but the result generalizes.)

In order to use Theorem 4.41 to exclude the exist-
ce of a (v,k,λ)-difference set in some group G, we first
ed to prove that the quadratic residues must necessarily be
ltipliers of such a difference set. The multiplier theorems
the next chapter will allow us to do this under certain
rcumstances. As a consequence, we will be able to obtain
e following corollary of Theorem 4.41.

Corollary 4.42. Suppose that there exists a (v,k,λ)-
fference set in an elementary abelian group G of order p^m,
ere p is a prime and $m \geq 2$. Suppose that

(1) $p=2q+1$ for some prime q and

(2) no prime divisor of n is congruent to 1 (mod p).

en n is not a prime.

Proof. Problem 7 of Chapter 5.□

For example, Corollary 4.4 rules out a (49,16,5)-
fference set in $\mathbb{Z}_7 \times \mathbb{Z}_7$ and a (529,33,2)-difference set in
$_{23} \times \mathbb{Z}_{23}$.

PROBLEMS - CHAPTER 4

1. Show that $x \mapsto x^{-1}$ is an automorphism of a group G if and only if G is abelian.

The next four problems generalize the difference set of Example 3. A <u>partial spread</u> of <u>cardinality</u> s in a group G of order $4N^2$ is a set of subgroups $H_1, \ldots, H_s$ of order 2N with the property that $H_i \cap H_j = \{1\}$ for $i \neq j$ (where 1 is the identity of G).

2. (i) Suppose that G has a partial spread $H_1, \ldots, H_N$ of cardinality N. Show that $(H_1 \cup \ldots \cup H_N) - \{1\}$ is a $(4N^2, 2N^2-N, N^2-N)$-difference set in G, closed under the permutation $g \mapsto g^{-1}$ (which is a multiplier if G is abelian).

(ii) Suppose that G has a partial spread $H_1, \ldots, H_{N+1}$ of cardinality N+1. Show that $(H_1 \cup \ldots \cup H_{N+1})$ is a $(4N^2, 2N^2+N, N^2+N)$-difference set in G, closed under $g \mapsto g^{-1}$.

3. (i) Suppose that a group H has automorphisms $\alpha_1, \ldots, \alpha_{N-2}$ with the property that $\alpha_i \alpha_j^{-1}$ fixes no nonidentity elements if $i \neq j$. Consider the following N subgroups of H x H: let $H_0 = \{(h,1) \mid h \in H\}$, $H_\infty = \{(1,h) \mid h \in H\}$ and $H_i = \{(h, \alpha_i(h)) \mid h \in H\}$ for $i=1,\ldots,N-2$. Show that $H_0, H_\infty, H_1, \ldots H_{N-2}$ is a partial spread in H x H.

(ii) Find a (36,15,6)-difference set in $\mathbb{Z}_6 \times \mathbb{Z}_6$.

(iii) Find a (36,15,6)-difference set in $S_3 \times S_3$, where S_3 is the symmetric group on 3 letters.

(iv) Find a $(2^{2t+2}, 2^{2t+1}-2^t, 2^{2t}-2^t)$-difference set in $(\mathbb{Z}_2)^{2t+2}$. (Let the automorphisms be powers of a single automorphism.)

(v) Find a $(4p^2, 2p^2-p, p^2-p)$-difference set in $D_{2p} \times D_{2p}$ where D_{2p} is the dihedral group of order 2p and p is a prime.

would be nice to obtain $(4N^2, 2N^2-N, N^2-N)$-difference sets abelian groups for $N \neq 2^t$ or 3 by using this construction. fortunately, the next problem shows that this is impossible.

Let G be an abelian group of order $4N^2$ containing a partial read $H_1, \ldots, H_N$. Count the number of elements of order 2 in and each H_i. Since no element of order 2 lies in more than e of the H_i, show that N=3 or $N=2^t$ for some integer t.

fact, we can show that an abelian 2-group with such a read is necessarily elementary abelian.

(i) Let $H_1, \ldots, H_r$ be a partial spread in an abelian group Show that $G \simeq H_i \times H_j$ for $i \neq j$. If $r \geq 3$ show that all the are isomorphic and that $G \simeq H_1 \times H_1$.

(ii) Show that every partial spread in an abelian group x H is equivalent (under an automorphism of H x H) to a read of the sort described in Problem 3.

(iii) Suppose that $|H| = 2^{t+1}$ and that H x H contains a rtial spread of cardinality 2^t. Suppose that H is the direct oduct of s cyclic groups. Count elements of order 2 to show at $s \geq t$. Thus, H is elementary abelian or $H \simeq \mathbb{Z}_4 \times (\mathbb{Z}_2)^{t-1}$. clude this case by observing that any automorphism of H xes a nonidentity element.

Construct a (36,15,6)-difference set in $\mathbb{Z}_4 \times \mathbb{Z}_3 \times \mathbb{Z}_3$. int: It's easier if you try to construct a "nice-looking" fference set.) There are two inequivalent answers to this oblem. Try to find both.

Find an example of a nontrivial difference set in an soluble group. (Hint: Let h be the order of your favourite mple group. Find a prime power q such that $q \equiv 1 \pmod{h}$. t d=h-2 in Example 4.)

Prove the assertion in Example $4\frac{1}{2}$. Let G be a group containing T as a central subgroup such that $G/T \simeq K$. Modify the proof of Theorem 1.12 to obtain an appropriate difference set in G. (Hint: Let $1=k_0, k_1, \ldots, k_r$ be coset representatives for T in G. For $i,j=0,1,\ldots,r$ define $\pi(i,j)$ by the relation $k_i k_j = Tk_{\pi(i,j)}$ and define t_{ij} by the equation $k_i k_j = t_{ij} k_{\pi(i,j)}$. Construct L as an $(r+1) \times (r+1)$ array of square matrices. In the i,j-th position put the matrix obtained by permuting the rows of $M_{\pi(i,j)}$ according to translation by t_{ij}.)

9. (i) Suppose that q and q+2 are prime powers. Let R_q and R_{q+2} be the Jacobsthal matrices of order q and q+2, respectively. Let

$$K = -(R_q \otimes R_{q+2}) - (I_q \otimes J_{q+2}) + (J_q \otimes I_{q+2}) + (I_q \otimes I_{q+2}).$$

Show that $KK^T = (q-1)^2 I - J$ and $KJ = JK = -J$.

(ii) Replace -1 by 0 in K and show that the resulting matrix is the incidence matrix of a Hadamard design.

(iii) Give a proof of Example 8.

10. Find a (133,33,8)-difference set by guessing a multiplier.

11. Consider H(q) as a difference set in the additive group of $\mathbb{F}_q$. Show that the transformations of the form $x \longmapsto u\alpha(x)$, where u is a nonzero square and α is an automorphism of $\mathbb{F}_q$, are multipliers of the difference set. Is this the full multiplier group? (You should not need to invoke Theorem 1.6.)

12. Generalize the "moreover" part of Theorem 4.5 as follows. Suppose that D is a symmetric (v,k,λ) design with a standard abelian automorphism group G having f+1 orbits on points, all but one of which consists of simply a fixed point. Suppose that for some prime p and integer j, we have $p^j \equiv -1$ (mod exponent of G). Show that p cannot divide n. (Theorem 4 is the case f=0.)

3. Suppose that -1 is a multiplier of a difference set in n abelian p-group. Show that p=2. (Hint: Use Theorem 4.14 nd Problems 3 and 4 of Chapter 2.)

4. (i) Show that for odd primes p, an integer t is a square mod p) if and only if t is a square (mod p^j) for all $j \geq 1$. What about p=2?)

(ii) Show that t is a square (mod ab) if and only if t is square (mod a) and (mod b), whenever a and b are relatively rime.

(iii) Hence, show that for an odd integer w, an integer t s a square (mod w) if and only if t is a square (mod p) for ll primes p dividing w.

. Let $\mathcal{D}$ be a nontrivial (v,k,λ)-difference set in an abelian oup G. Prove that if $\mathcal{D}$ admits -1 as a multiplier then G ntains an elementary abelian 2-group of order at least + $k-1/\lambda$, and at least $1 + k/\lambda$ if k and λ are both even.

What information does this give about a (4000,775,150)-fference set admitting -1 as a multiplier?

. Generalize Theorem 4.25 as follows. Let $\mathcal{D}$ be a (v,k,λ)-fference set in an abelian group G. Suppose that G = PxQ ere $(|P|,|Q|)=1$ and every prime dividing $|P|$ also divides If every prime dividing $|P|$ is self-conjugage (mod $|Q|$) en for some prime p dividing $|P|$ and some integer i:

$$p^{2i} \| n, \quad p^i \| k, \quad p^i \| \lambda \quad \text{and} \quad p^{i+1} | v.$$

heorem 4.25 is the case in which $|P|$ has only one prime ctor.)

t H be an abelian group. A group G is called a <u>generalized</u> <u>nedral extension</u> of H if G is generated by H together with element q such that $q^2=1$ and $qhq=h^{-1}$ for all $h \in H$.

17. Suppose that a generalized dihedral extension G of an abelian group H contains a (v,k,λ)-difference set $\mathbb{D}$. Show that any abelian group K containing H as a subgroup of index 2 also contains a (v,k,λ)-difference set. (Hint: Let $\mathbb{D}=\mathbb{D}_1 \cup \mathbb{D}_2 q$ where $\mathbb{D}_1, \mathbb{D}_2 \subseteq H$. Suppose that $K=H \cup H\theta$. Set $\mathbb{D}'=\mathbb{D}_1 \cup \mathbb{D}_2\theta$.)

18. Use the previous problem and the fact that $\mathbb{Z}_{16}$ does not contain a (16,6,2)-difference set to show that the dihedral group of order 16 also does not.

There are fourteen groups of order 16, up to isomorphism. (See, e.g., Burnside [23].) With the exception of the cyclic and dihedral groups, the other twelve possess (16,6,2)-difference sets. What is most amazing is that each of these twelve groups can be realized as a regular automorphism group of the single design $S^-(4)$. The next problem shows that one of these groups, $\mathbb{Z}_2 \times \mathbb{Z}_8$ can also be realized as a regular automorphism group of a different symmetric (16,6,2) design.

19. Consider the sets $\mathbb{D}_1=\{(0,0),(0,1),(0,2),(0,5),(1,0),(1,6$ and $\mathbb{D}_2=\{(0,0),(1,0),(0,1),(1,2),(1,5),(1,6)\}$ in $\mathbb{Z}_2 \times \mathbb{Z}_8$. Show that both are (16,6,2)-difference sets, that $\mathrm{dev}(\mathbb{D}_1) \simeq S^-($ and that $\mathrm{dev}\ (\mathbb{D}_2) \not\simeq S^-(4)$. (Hint: You have seen the incidence matrices before.)

20. Suppose that $\mathbb{D}$ is a $(4p^2, 2p^2 \pm p, p^2 \pm p)$-difference set in an abelian group G, with p a prime congruent to 3 (mod 4). We show that p=3.

(i) Let V be the Sylow 2-subgroup of G, let P_1 be a subgroup of order p and let P be the Sylow p-subgroup of G. Let A be the incidence matrix of dev $(\mathbb{D})$. Determine the entries of A_P.

(ii) Consider A_{P_1}. For any particular row, show that entries corresponding to columns in the same coset of V are congruent (mod p). Show that, for at least one coset, the columns corresponding to this coset cannot all be equal. (Hint: See the proofs of Lemmas 4.35, 4.36, and 4.37.) Determine the possibilities for the entries in these columns.

(iii) Use the fact that A_p is a contraction of A_{p_1} to show that p=3.

21. Show that the preceding problem remains valid for $p \equiv 1 \pmod 4$ provided that $V = \mathbb{Z}_2 \times \mathbb{Z}_2$. What if $V = \mathbb{Z}_4$?

22. Show that there does not exist an (81,16,3)-difference set $\mathbb{D}$ in $(\mathbb{Z}_3)^4$. (Hint: Regard $(\mathbb{Z}_3)^4$ as an affine space over $\mathbb{Z}_3$. Show by contraction that the translates of any 3-dimensional subspace must have 3, 6 and 7 points of $\mathbb{D}$, respectively. Let H be a translate containing exactly 3 points of $\mathbb{D}$. Consider the four translates (including H) of 3-dimensional subspaces containing these 3 points. How many points of $\mathbb{D}$ does each contain?)

23. Suppose that D is a (v,k,λ)-difference set in an abelian group G, with -1 as a multiplier. Suppose that $p^{2i} \| n$ and $p^{i+e} \| v$, for some prime p and positive integers i and e. Show that the Sylow p-subgroup of G has exponent at most p^e. (Hint: Use the technique of §4.7.)

A <u>polarity</u> is a self-inverse automorphism from a symmetric design to its complement (cf. §1.3.). We say that p is n <u>absolute point</u> of a polarity σ if p is incident with σp.

4. Let $\mathbb{D}$ be a (v,k,λ)-difference set in an abelian group G. he map $x \longmapsto \mathbb{D}-x$ defines a polarity of dev(D). The number ρ f absolute points is the number of solutions of $2x \in D$. This umber can vary between different translates of $\mathbb{D}$. However, rove:

(i) ρ is always divisible by the order of the (unique) argest subgroup of G of exponent 2;

(ii) the average value of ρ (taken over all translates) s k;

(iii) $\rho=k+g\sqrt{n}$, for some integer $g\equiv v-1 \pmod 2$. (Hint: write the incidence matrix symmetrically, and consider its eigenvalues and trace.)

25. Show that there does not exist a (40,13,4)-difference set in $(\mathbb{Z}_2)^3 \times \mathbb{Z}_5$. (Hint: Use the previous problem. Show that $\rho=13+3g\equiv 0 \pmod 8$, so that $\rho=16$ or 40.)

26. Use problem 24 to investigate a (352,27,2)-difference set (cf. §5.4).

NOTES TO CHAPTER 4

§4.1 The classical papers on difference sets are nger [126] and Hall [48]. Example 4 is due to McFarland [97] though our treatment follows Lenz and Jungnickel [83]. ample $4\frac{1}{2}$ is due to Dillon [36]. Concerning Example 6, the gular automorphism groups acting on PG(m,q) have all been termined. (See [33, p. 35].) In particular there is always nonabelian regular group if the prime power q is not a prime. ncerning Example 7, the extensive work done on n-th power sidues is surveyed in Baumert [13]. The twin prime difrence sets in Example 9 are due to Stanton and Sprott [129]. e reader may also be interested in the work of Gordon, Mills d Welch [43] who constructed cyclic difference sets with the me parameters as PG(m,q) but which are not equivalent to e Singer-cycle difference sets. It is unknown whether the mmetric designs associated with these difference sets are, general, not isomorphic to the corresponding PG(m,q). 'his is the case for the first few cases of the construction, least.)

Multipliers seem to have been first defined by Hall 48].

Theorem 4.4 has been known for abelian groups G for me time. See [91]. The proof in the text, however, is the rst to include nonabelian groups as well.

§4.2 Mann [91] undertook one of the first systemic investigations of abelian difference sets; he proved eorem 4.5 by different methods.

Kantor [72] proved Theorem 4.11 using a much more mplicated counting argument. The present proof is due to nder [79].

§4.3 Theorems 4.13, 4.14 and 4.15 can all be found Mann [91].

§4.4 The results of this section are due to Lander (although Theorem 4.18 also follows as a special case of a result of Turyn [132]).

§4.5 Difference sets with multiplier -1 were first systematically investigated by Johnsen [70] who provided Corollaries 4.22 and 4.24. Johnsen also proved a weaker version of Theorem 4.25 (which asserts only that $p^i | v$). The stronger version in the text and Theorem 4.27 are proved by Lander [79]. Theorem 4.28 follows from McFarland and Rice [99].

The (4000,775,150)-difference set in $(\mathbb{Z}_5)^3 \times (\mathbb{Z}_2)^5$ with multiplier -1 is due to McFarland, who notes that the equation $q^d+\ldots+q+2=2^u$ has no further solutions in prime powers q when d=2.

Suppose that D is a (v,k,λ)-difference set in G admitting -1 as a multiplier. The map $x \mapsto D+x$ is a polarity, for which every point is absolute (if $0 \in D$) or no point is absolute (if $0 \notin D$). The incidence matrix of dev(D) can then be written as a symmetric matrix with constant diagonal. Ignoring the diagonal entries, it defines a graph Γ with valency k-1 or k, respectively. In fact, Γ is a strongly regular graph admitting G as a regular automorphism group. (Concerning the interesting topic of strongly regular graphs, see [25].) We should remark that all of the methods in Chapter 4 carry over straightforwardly to the general study of strongly regular graphs with regular abelian automorphism groups.

§4.6 Theorem 4.30 and 4.31 also follow from Turyn [132], although the proofs here are much different and rather simpler. Theorem 4.32 is due to Lander.

§4.7 This section is due to Turyn [132].

§4.8 Further results on difference sets are found in Baumert [12].

Problems. Problems 2, 4 and 4 are taken from Dillon
]. Problems 12 and 16 are due to Lander. Problem 17 is
to Dillon. For a reference on (16,6,2)-difference sets,
[9] and [36]. Problems 20 and 21 are taken from Mann
McFarland [95]. Problem 22 is due to Gleason and is pub-
hed in Turyn [132]. Problem 23 is due to Lander [79].

5. MULTIPLIER THEOREMS

§5.1 THE AUTOMORPHISM THEOREM

An automorphism of a symmetric design D naturally induces an automorphism of the R-module spanned by the rows of the incidence matrix of D (where R is any ring). We explore in this section a converse question. Suppose that a permutation π of the points of D induces an automorphism of certain of the R-modules. When can we conclude that π must actually be an automorphism of D?

The following result provides an answer. (Some notation: if N is a $\mathbb{Z}$-module in $\mathbb{Z}^v$, let $N_{(\text{mod } m)}$ be the image of N under the "reduction modulo m" homomorphism of $\mathbb{Z}^v$.)

<u>Theorem 5.1.</u> <u>Let</u> D <u>be a symmetric</u> (v,k,λ) <u>design and let</u> M <u>be the</u> $\mathbb{Z}$<u>-module of</u> D. <u>Let</u> n_1 <u>be a divisor of</u> n. <u>Suppose that</u> π <u>is a permutation of the points of</u> D. <u>If</u>: (1) π <u>induces an automorphism of</u> $M_{(\text{mod } n_1)}$, <u>and</u> (2) $n_1 > \lambda$

<u>then</u> π <u>defines an automorphism of</u> D.

It is convenient to slightly rephrase Theorem 5.1. Suppose that $n_1 = p_1^{\alpha_1} \ldots p_s^{\alpha_s}$ is the canonical prime factorization of n_1. Then condition (1) holds if and only if π induces an automorphism of $M_{(\text{mod } p_i^{\alpha_i})}$, for $i=1,\ldots,s$. So

<u>Theorem 5.1 (The Automorphism Theorem)</u> <u>Let</u> D <u>be a symmetric</u> (v,k,λ) <u>design and let</u> M <u>be the</u> $\mathbb{Z}$<u>-module of</u> D. <u>Let</u> n_1 <u>be a divisor of</u> n <u>with canonical prime factorization</u> $n_1 = p_1^{\alpha_1} \ldots p_s^{\alpha_s}$.

<u>Suppose that</u> π <u>is a permutation of the points of</u> D. <u>If</u>: (1) π <u>induces an automorphism of</u> $M_{(\mathrm{mod}\ p_i^{\alpha_i})}$,

<u>for</u> i=1,...,s, <u>and</u>

(2) $n_1 > \lambda$

en π <u>defines an automorphism of</u> D.

<u>First proof</u>. Let A be the incidence matrix of D.

e module M is the $\mathbb{Z}$-span of the rows of A. The extended dule M^{ext} is the $\mathbb{Z}$-span of the rows of

$$B = \left(\begin{array}{c|c} & 1 \\ A & \vdots \\ & 1 \\ \hline \lambda \ldots \lambda & k \end{array}\right)$$

make two observations. If $x = (x_1,\ldots,x_{v+1})$ and $y =$ $_1,\ldots,y_{v+1})$ are elements of M^{ext} then

(1). $x_1 + \ldots + x_v = kx_{v+1}$ (mod n) and

(2) $\psi(x,y) \equiv 0$ (mod n), where

$$\psi(x,y) = x_1y_1 + \ldots + x_vy_v - \lambda x_{v+1}y_{v+1}.$$

Now, let E be an arbitrary block of D and write

$$(\ldots\ E\ \ldots)$$

an abbreviation for the row of A that is the characteristic nction of E. Let E^π be the image of the block E under π. x an integer i with $1 \le i \le s$. Since π defines an automorphism f $M_{(\mathrm{mod}\ p_i^{\alpha_i})}$, then

$$(\ldots\ E^\pi\ \ldots) \in M_{(\mathrm{mod}\ p_i^{\alpha_i})}.$$

, we can write this vector as a linear combination od $p_i^{\alpha_i}$) of the rows of A. If we take precisely the same linear combination of the corresponding rows of the extended matrix B, we obtain a vector

$$x = (\ldots E^{\pi} \ldots | u) \in M^{ext} \pmod{p_i^{\alpha_i}}$$

where u is some integer. While we do not know what integer u is, we know that $k = ku \pmod{p_i^{\alpha_i}}$, by the first observation above. Since $k \equiv \lambda \pmod{p_i^{\alpha_i}}$, then also $\lambda \equiv \lambda u \pmod{p_i^{\alpha_i}}$.

Let F be an arbitrary block of D (possibly the same as E) and let

$$y = (\ldots F \ldots | 1),$$

an element of $M^{ext} \pmod{p_i^{\alpha_i}}$. Now, $\psi(x,y) \equiv 0 \pmod{p_i^{\alpha_i}}$, by the second observation above. Hence

$$\begin{aligned} 0 &\equiv \psi(x,y) && \pmod{p_i^{\alpha_i}} \\ &\equiv |E^{\pi} \cap F| - \lambda u && \pmod{p_i^{\alpha_i}} \\ &\equiv |E^{\pi} \cap F| - \lambda && \pmod{p_i^{\alpha_i}}. \end{aligned}$$

Thus,

$$|E^{\pi} \cap F| \equiv \lambda \pmod{p_i^{\alpha_i}}.$$

Since this holds for $i=1,\ldots,s$, then

$$|E^{\pi} \cap F| \equiv \lambda \pmod{n_1}.$$

Finally, since $n_1 > \lambda$ then $|E^{\pi} \cap F| \geq \lambda$. Then E^{π} meets every block in at least λ points. By the lemma below, E^{π} must be a block of D. Since E was an arbitrarily chosen block, then π induces an automorphism of D.

Lemma 5.2. Let D be a symmetric (v,k,λ) design. Let S be a set of k points which intersects every block of D in at least λ points. Then S is a block.

Proof of lemma. Let $B_1,\ldots,B_v$ be the blocks of D.
$t_i = |S \cap B_i| - \lambda$, which is a nonnegative integer by
othesis. Then

$$\sum_{i=1}^{v} t_i = \sum_{i=1}^{v} (|S \cap B_i| - \lambda) = k^2 - v\lambda = n.$$

ondly, by counting in two ways the number of triples
x,y) with B a block and x,y distinct points of $B \cap S$, we
that

$$\sum_{i=1}^{v} (t_i + \lambda)(t_i + \lambda - 1) = k(k-1)\lambda.$$

ng the equation $\sum t_i = n$, this equation reduces to

$$\sum_{i=1}^{v} t_i^2 = n^2$$

ce the t_i are nonnegative, we must have for some integer j

$$t_i = \begin{cases} n & \text{if } i=j, \\ 0 & \text{otherwise.} \end{cases}$$

s $|S \cap B_j| = k$ and $S = B_j$.

This proves the theorem.□

Remark. While the proof is quite transparent, it
unfortunately not the best proof from the point of view of
eralizations.

In the next section, we shall give a slightly dif-
ent approach which will in turn point the way to further
eralizations.

In order to make use of the Automorphism Theorem, we must know automorphism of the modules $M_{(\text{mod } p_i^{\alpha_i})}$.

Unfortunately, for an arbitrary symmetric design D, there is little we can say about these modules and their automorphisms. In the case of difference sets in a group G, however, the modules are ideals in a group ring. We have already noticed that there are certain mappings which are automorphisms of _every_ ideal.

Fact 1. _Let_ G _be an abelian group of order_ v _and let_ p _be a prime not dividing_ v. _Then the map_ $\alpha : \mathbb{F}_p G \longrightarrow \mathbb{F}_p G$ _given by_

$$\alpha(x) = x^{(p)}$$

for $x \in \mathbb{F}_p G$ _is an automorphism of_ $\mathbb{F}_p G$ _fixing every ideal of_ $\mathbb{F}_p G$.

That is, p _is a multiplier of every ideal of_ $\mathbb{F}_p G$.

Fact 1 is simply Proposition 4.7 restated. Combining this fact and the automorphism theorem, we can prove a theorem about abelian difference sets.

Theorem 5.3.(First Multiplier Theorem). _Let_ $\mathcal{D}$ _be a_ (v,k,λ)-_difference set in an abelian group_ G. _Let_ p _be a prime which divides_ n _but does not divide_ y.

_If_ $p>\lambda$ _then_ p _is a multiplier of_ $\mathcal{D}$.

Proof. Let D be the development of $\mathcal{D}$. Apply the Automorphism Theorem with $n_1=p$. By Fact 1, the module $M_{(\text{mod } p)}$--that is, the $\mathbb{F}_p$-code--admits p as a multiplier.

Hence, p is a multiplier of $\mathcal{D}$.□

Theorem 5.3 is extremely useful for investigating putative difference sets. For example, suppose that $\mathcal{D}$ is a (37,9,2)-difference set in $\mathbb{Z}_{37}$. By the First Multiplier Theorem, 7 is a multiplier. Let $\mathcal{D}'$ be a translate of $\mathcal{D}$ fixed by this multiplier and let a be a nonzero element of $\mathcal{D}'$. Then $a, 7a, 7^2a, \ldots \in \mathcal{D}'$. Thus {a,7a,12a,10a,33a,9a,25a,34a,16a} $\subseteq \mathcal{D}'$. This accounts for all nine elements of $\mathcal{D}'$. A quick check confirms that this set is indeed a (37,9,2)-difference set. So, starting only with knowledge of the multiplier 7, have constructed such a difference set and seen that it is que up to equivalence.

The First Multiplier Theorem is also a powerful tool proving nonexistence. We mention two examples:

(1) Suppose that there exists a (79,13,2)-difference D in $\mathbb{Z}_{79}$. Necessarily, 11 is a multiplier. Let D' be a nslate fixed by this multiplier. Then D' is a set of 13 ments closed under multiplication by 11. But, the "multi- cation by 11" map on $\mathbb{Z}_{79}$ has three orbits of sizes 1,39 l 39, respectively (since 11 has order 39 (mod 79)). The tradiction shows that no such difference set exists.

(2) Suppose that there exists a (529,33,2)-difference D in $\mathbb{Z}_{529}$. Then 31 is a multiplier. Since "multiplica- on by 31" has orbits of sizes 1,11,11,253 and 253, we can d no translate of D fixed by the multiplier. Hence no such fference set exists.

Table 5-1 lists examples of difference sets excluded such an argument.

ble 5-1

,k,λ)	n	Group	Multiplier	Orbit Sizes
31,10,3)	7	(31)	7	{1,15,15}
79,13,2)	11	(79)	11	{1,39,39}
49,16,15)	11	(49)	11	{1,3,3,21,21}
69,17,4)	13	(3)(23)	13	{1,1,1,1,11,11,11,11,11,11}
191,20,2)	18	(191)	3	{1,95,95}
211,21,2)	19	(211)	19	{1,15,15,...,15}
301,25,2)	23	(7)(43)	23	{1,3,3,23,23,69,69,69,69}
131,26,5)	21	(131)	7	{1,65,65}
127,28,6)	22	(127)	11	{1,63,63}
813,29,1)	28	(3)(271)	2	{1,2,135,135,270,270}

Table 5-1 (continued)

(v,k,λ)	n	Group	Multiplier	Orbit Sizes
(529,33,2)	31	(529)	31	{1,11,11,253,253}
(631,36,2)	34	(631)	17	{1,315,315}
(421,36,3)	33	(421)	11	{1,105,105,105,105}
(253,36,5)	31	(11)(23)	31	{1,5,5,11,11,55,55,55,55}
(181,36,7)	29	(181)	29	{1,15,15,...,15}
(223,37,6)	31	(223)	31	{1,111,111}
(991,45,2)	43	(991)	43	{1,495,495}
(691,46,3)	43	(691)	43	{1,345,345}
(139,46,15)	31	(139)	31	{1,69,69}
(1171,49,2)	47	(1171)	47	{1,195,195,...,195}
(393,49,6)	43	(3)(131)	43	{1,1,1,65,65,...,65}
(295,49,8)	41	(5)(59)	41	{1,1,1,1,1,29,29,...,29}

We can improve the First Multiplier Theorem if we make use of a result stronger than Fact 1.

Fact 2. Let G be an abelian group of order v and let p be a prime not dividing v. Then the map $\alpha : \hat{\mathbb{Z}}_p G \longrightarrow \hat{\mathbb{Z}}_p$ given by

$$\alpha(x) = x^{(p)}$$

for $x \in \hat{\mathbb{Z}}_p G$ is an automorphism of $\hat{\mathbb{Z}}_p G$ fixing every ideal of $\hat{\mathbb{Z}}_p G$.

That is, p is a multiplier of every ideal of $\mathbb{F}_p G$.

The ring $\hat{\mathbb{Z}}_p$ is the ring of p-adic integers. It is discussed and Fact 2 is proven in Appendix F. The advantage of working with $\hat{\mathbb{Z}}_p$ is that the ring $(\mathbb{Z}/p^\alpha\mathbb{Z})$, of integers (mod p^α), is a homomorphic image of $\hat{\mathbb{Z}}_p$ for all $\alpha \geq 1$. Thus, Fact 2 entails the analogous statement for all the rings $(\mathbb{Z}/p^\alpha\mathbb{Z})$. Using this in conjunction with the Automorphism Theorem, we obtain the following result:

Theorem 5.4. (Second Multiplier Theorem). Let $\mathcal{D}$ be a (v,k,λ)-difference set in an abelian group G. Let n_1 be a divisor of n and let $n_1 = p_1^{\alpha_1} \ldots p_s^{\alpha_s}$ be its canonical prime factorization.

Suppose that t is an integer relatively prime to v. If: (1) for $i=1,\ldots,s$, there exists an integer $j=j(i)$ such that $t = p_i^{\,j}$ (mod exponent of G), and

(2) $n_1 > \lambda$

then t is a multiplier of $\mathcal{D}$.

Proof. Let D be the development of $\mathcal{D}$. Consider the point permutation given by the map $\gamma: g \longmapsto g^t$ for $g \in G$. (Since $(t,v)=1$, this is indeed an automorphism of G.)

By Fact 2, the map $\gamma_i: g \longmapsto g^{p_i}$ is an automorphism of the module $M_{(\text{mod } p_i^{\alpha_i})}$. Hence $\gamma = \gamma_i^{\,j(i)}$ is an automorphism of the module $M_{(\text{mod } p_i^{\alpha_i})}$. By the Automorphism Theorem, γ defines an automorphism of D.□

We mention two applications of the Second Multiplier Theorem:

(1) Suppose that there exists a (71,15,3)-difference set $\mathcal{D}$ in $\mathbb{Z}_{71}$. Taking $n_1=4$, we see that 2 must be a multiplier. However, since "multiplication by 2" has orbits of size 1,35 and 35, no translate of $\mathcal{D}$ could be fixed by this multiplier. Hence, no such difference set exists.

(2) Suppose that there exists a (239,35,5)-difference set $\mathcal{D}$ in $\mathbb{Z}_{239}$. Let $n_1=6$ and $t=2$. The hypotheses of Theorem 5.4 are satisfied since $t \equiv 2$ (mod 239) and $t \equiv 3^{62}$ (mod 239). Thus, 2 is a multiplier. Since "multiplication by 2" has orbits of size 1,119 and 119, no translate could be fixed by this multiplier. Hence no such difference set exists.

Table 5-2 lists examples of difference sets excluded by such an argument.

TABLE 5-2

(v,k,λ)	n	Group	Multiplier	Orbit Sizes
(71,15,3)	12	(71)	2	{1,35,35}
(71,21,6)	15	(71)	$3 \equiv 3^1$ $\equiv 2^{16}$	{1,35,35}
(311,31,3)	28	(311)	2	{1,155,155}
(239,35,5)	30	(239)	$2 \equiv 2^1$ $\equiv 3^{62}$	{1,119,119}
(149,37,9)	28	(149)	$7 \equiv 7^1$ $\equiv 2^{142}$	{1,74,74}
(391,40,4)	36	(17)(23)	3	{1,11,11,16,176,176}
(329,41,5)	36	(7)(47)	3	{1,3,3,23,23,69,69,69,69}

The argument above works whenever there is no way of forming a union of orbits having cardinality k. Whenever there is at least one such way--but not too many ways--we may settle the existence question directly by trying each way to see if it yields a difference set.

§5.2 CONTRACTED AUTOMORPHISM THEOREM

With a few slight modifications the Automorphism Theorem applies to the contraction of a symmetric design by a semi-regular automorphism group.

Theorem 5.5. (Contracted Automorphism Theorem). *Let* D *be a symmetric* (v,k,λ) *design with a semi-regular automorphism group* H, *having order* h *and* w *orbits. Let* M^H *be the* $\mathbb{Z}$*-module spanned by the rows of the contracted incidence matrix* A_H.

Let n_1 *be a divisor of* n *with canonical prime fac-*
rization $n_1 = p_1^{\alpha_1} \ldots p_s^{\alpha_s}$.

Suppose that π *is a permutation of the points of* D_H.
If: (1) π *induces an automorphism of* $M^H \pmod{p_i^{\alpha_i}}$,

for i=1,...,s, *and*

(2) *either* $n_1 > h\lambda$ *or* $n_1 = n$,

en π *defines an automorphism of* D_H.

Proof. We shall take a slightly different approach
an we did in the proof of Theorem 5.1--one which will point
e way to further generalizations in later sections.

Let $(A_H)^\pi$ be the matrix obtained by permuting the
lumns of A_H according to π. We must show that $(A_H)^\pi$ has
e same rows as A_H, although possibly in a different order.
nsider the matrix

$$E = \left(\begin{array}{c|c} A_H & \begin{matrix} 1 \\ \cdot \\ \cdot \\ \cdot \end{matrix} \\ \hline (A_H)^\pi & \begin{matrix} 1 \\ \cdot \\ \cdot \\ \cdot \\ 1 \end{matrix} \end{array}\right)$$

garding rows as vectors in $\mathbb{Z}^{w+1} \subseteq \mathbb{Q}^{w+1}$ and using the scalar
oduct $\psi(x,y) = x_1y_1 + \ldots + x_wy_w - h\lambda x_{w+1}y_{w+1}$, we compute the matrix
inner products

$$F = E\psi E^T = \left(\begin{array}{c|c} nI & S^T \\ \hline S & nI \end{array}\right)_{2w \times 2w}$$

ere $S = (A_H)^\pi A_H{}^T - h\lambda J$.

Now, π induces an automorphism of D_H if and only if
$A_H)^\pi = A_H$ for some permutation matrix P. Equivalently,
duces an automorphism of D_H if and only if

$$\begin{aligned} PS &= P[(A_H)^{\pi} A_H^T - h\lambda J] \\ &= A_H A_H^T - h\lambda J \\ &= nI . \end{aligned}$$

--that is, if and only if S is n times a permutation matrix.

Notice that the rank of E as a rational matrix is w. (It is at least w since A_H is nonsingular and at most w since the sum of the first w columns is a multiple of the last.) Hence F also has rank w. Therefore

$$SS^T = S^T S = n^2 I.$$

(To see this, notice that if we use the upper left block of E to clear the lower left by row elimination, then we obtain $nI - n^{-1}SS^T$ in the lower right. Since rank F=w, this must be zero.) Secondly notice that

$$\begin{aligned} SJ = JS &= [(A_H)^{\pi} A_H^T - h\lambda J]J \\ &= (k^2 - v\lambda)J \\ &= nJ. \end{aligned}$$

Notice that we have not made use of conditions (1) and (2) in proving that $SS^T = n^2I$ and $SJ = JS = nJ$.

We shall use these conditions now to show that S must be n times a permutation matrix. Proceed now as in the proof of Theorem 5.1. Let $e_1,\ldots,e_w,f_1,\ldots,f_w$ be the rows of E. Since π induces an automorphism of $M^H{}_{(\mathrm{mod}\ p_i^{\alpha_i})}$ then all of these vectors lie in $M^H{}_{(\mathrm{mod}\ p_i^{\alpha_i})}$. Hence

$$\psi(e_a, f_b) \equiv 0 \qquad (\mathrm{mod}\ p_i^{\alpha_i})$$

for $1 \le a, b \le w$. Hence $S \equiv 0\ (\mathrm{mod}\ p_i^{\alpha_i})$, for $i=1,\ldots,s$. Thus,

$$S \equiv 0 \pmod{n_1}.$$

Set $T = \frac{1}{n_1} S$. Then T is an integral matrix such that

$$TT^T = (n/n_1)^2 I$$

and $$TJ = JT = (n/n_1)J.$$

Moreover, since the entries of S are greater than or equal to $-h\lambda$, then the entries of T are integers greater than or equal to $-h\lambda/n_1$.

<u>Case 1</u>. $n_1 > h\lambda$. In this case the entries of T are nonnegative integers. If $(t_1,\ldots,t_w)$ is a row or column of T then

$$\sum t_i^2 = (n/n_1)^2 \quad \text{and} \quad \sum t_i = (n/n_1).$$

All the t_i must be zero except for one, which must be n/n_1. Since this is true for any row or column, T must be a multiple of a permutation matrix.

<u>Case 2</u>. $n_1 = n$. In this case, $TT^T = I$. Any such integral matrix must be a permutation matrix. (Why?)

In both cases S is a multiple of a permutation matrix, which proves the theorem.

<u>Remarks</u>. (1) The key to the proof is the matrix equations $SS^T = n^2 I$ and $SJ = JS = nJ$. The hypotheses allow us to conclude that S is a multiple of a permutation matrix and thus that π is an automorphism. In §§5.4 and 5.5, we will generalize the method by finding other hypotheses that allow us to reach the same conclusion about S.

(2) The reader may wonder why we did not include both of the conditions "$n_1 > \lambda$" and "$n_1 = n$" in Theorem 5.1, which is merely the special case of h=1 in Theorem 5.5. The reason is that, after possibly replacing a symmetric design by its complement, we have $n > \lambda$. Hence the latter condition is superfluous.

Using Theorem 5.5 in conjunction with Fact 2, we obtain a contracted multiplier theorem.

Theorem 5.6. (Contracted Multiplier Theorem) Let $\mathcal{D}$ be a (v,k,λ)-difference set in an abelian group G. Let n_1 be a divisor of n and let $n_1=p_1^{\alpha_1}\ldots p_s^{\alpha_s}$ be its canonical prime factorization.

Suppose that H is a subgroup of order h and index w. Suppose that t is an integer relatively prime to w.

If: (1) for $i=1,\ldots,s$, there exists an integer $j=j(i)$ such that $t\equiv p_i^{\,j}$ (mod exponent of G/H), and

(2) either $n_1>h\lambda$ or $n_1=n$

then t is a G/H-multiplier.

We mention some applications of the Contracted Multiplier Theorem:

(1) Suppose that $\mathcal{D}$ is a (141,36,9)-difference set in the abelian group $G=\mathbb{Z}_3 \times \mathbb{Z}_{47}$. Let H be a subgroup of order 3. By the theorem above, 3 must be a G/H-multiplier. Since this multiplier fixes a column of the contracted incidence matrix A_H, it also fixes a row of A_H. Consider such a row. The order of 3 (mod 47) is 23. Thus the G/H-multiplier has orbits or sizes 1,23 and 23. The fixed row has the form

$$(a,\underbrace{b,\ldots,b}_{23},\underbrace{c,\ldots,c}_{23})$$

Since

$$A_H A_H^T = 27\ I + 9\cdot 3\ J$$

and

$$A_H J = JA_H = 36J$$

we have

$$a^2 + 23b^2 + 23c^2 = 54$$

and

$$a+ 23b + 23c = 36.$$

'hese equations have no common integral solution. (In fact, he first has no integral solutions at all.) Consequently o such difference set exists.

(2) Suppose that $\mathcal{D}$ is a (177,33,6)-difference set n the abelian group $G=\mathbb{Z}_3 \times \mathbb{Z}_{59}$. Let H be a subgroup of rder 3. Again, 3 is a G/H-multiplier. As before, consider row of the contracted incidence matrix A_H which is fixed by his multiplier. Since the order of 3 (mod 59) is 29, this ow has the form

$$(a,\underbrace{b,\ldots,b}_{29},\underbrace{c,\ldots,c}_{29}).$$

ince

$$A_H A_H^T = 27\ I + 6\cdot 3\ J$$

and

$$A_H J = J A_H = 33,$$

e have

$$a^2 + 29b^2 + 29c^2 = 45$$

and

$$a + 29b + 29c = 33.$$

'hese equations have (4,1,0) and (4,0,1) as their only non-egative integral solutions. However, by the construction of $_H$, all entries of this matrix are at most 3. Consequently, o (177,33,6)-difference set in G can exist.

5.3 BLOCKS FIXED BY MULTIPLIERS

When using a multiplier α to study a putative dif-'erence set, we begin by taking a block fixed by α. Propo-ition 3.1 assures us that there is such a block. In this ection, we ask more generally: if Γ is a group of multipliers f an abelian difference set, must there exist a block fixed y every element of Γ?

If Γ is cyclic the answer is Yes (by Proposition 3.1). In general, however, the answer is No. As a counter-example, consider the difference set in $(\mathbb{Z}_2)^{2m}$ defined by $S^{\pm}(2m)$. The full group of multipliers is transitive on nonidentity elements. By Proposition 3.1, the group Γ has two block orbits--which must be the blocks containing the identity and the blocks not containing the identity. So, Γ fixes no block.

If we are to modify our question to obtain a Yes answer, we must either place restrictions on the difference set or on the group Γ. We explore several such theorems.

Theorem 5.7. Let $\mathcal{D}$ be a (v,k,λ)-difference set in an abelian group G. The group of numerical multipliers fixes at least one translate of $\mathcal{D}$.

Proof. Let D be the development of $\mathcal{D}$. Write G as a direct product of cyclic groups of prime power order. Say, $G=G_1 \times \ldots \times G_s$. Let $H_j=G_1 \times \ldots \times G_{j-1} \times G_{j+1} \times \ldots \times G_s$, for $j=1,\ldots,s$. Let M be the group of all numerical multipliers of $\mathcal{D}$. We observed in Chapter 4 that M induces a group M_j of multipliers in the contraction D_{H_j}.

Now, the cyclic group G_j acts regularly on D_{H_j}. We show first of all that some block of D_{H_j} is fixed by all multipliers in M_j. This is trivial if M_j is cyclic, for we simply take a block fixed by a generator of M_j. What if M_j is not cyclic? Well, the full automorphism group A_i of a cyclic group G_i of prime power order t is isomorphic to the multiplicative group of units (mod t). By Problems 8 and 9, the group A_i is cyclic unless $t=2^e$ with $e\geq 3$. In this case, the group is a direct product of a group of order 2 and a group of order 2^{e-2}. Furthermore, any noncyclic subgroup of A_i must contain all three elements of order 2. In particular, $\alpha: g \longmapsto g^{-1}$ must be contained in any noncyclic group of automorphisms. Thus if M_j is not cyclic, then $|G_i|$ is a power of 2 and -1 is a G/H_j-multiplier. Thus, v is even and n is a square. By the remark following Theorem 4.28, there exists a block of D_{H_j} fixed by every automorphism in M_j.

The blocks of D_{H_j} are obtained by summing the rows as g ranges over a coset bH_j of H_j. For j=1,...,s, let be an element of G such that b_jH_j corresponds to a block D_{H_j} fixed by all numerical multipliers. Then the set

$$\{gD \mid g \in b_jH_j\}$$

fixed setwise by all numerical multipliers. Thus the tersection

$$\bigcap_{j=1}^{s} \{gD \mid g \in b_jH_j\} = \{gD \mid g \in \bigcap_{j=1}^{s} b_jH_j\}$$

fixed setwise by all numerical multipliers. But

$\bigcap_{j=1}^{s} b_jH_j| = |\bigcap_{j=1}^{s} H_j| = 1$. Thus the intersection contains ecisely one translate, which is then fixed by all multipliers.□

Remark. The proof fails for non-numerical multi- iers α since in general $\alpha(H_j) \neq H_j$ and thus α does not induce multiplier of D_{H_j}.

Theorem 5.7 restricts the group Γ. The next two eorems allow any multiplier group Γ but place restrictions D.

Lemma 5.8. Let J and K be permutation groups on a t X. If jk=kj for all $j \in J$ and $k \in K$, then J permutes etwise) the objects fixed by K.

In particular, if K fixes exactly one element x, en J also fixes x.

Proof. Let K fix x. Then k(jx)=j(kx)=jx, for all $\in K$. Hence K fixes jx.□

Theorem 5.9. Let D be a (v,k,λ)-difference set in abelian group G. Suppose that $t_1,\ldots,t_s$ are numerical ltipliers of D.

If $(v,t_1-1,\ldots,t_s-1)=1$, *then there is a unique translate of* $\mathcal{D}$ *fixed by all multipliers of* $\mathcal{D}$.

Proof. By Theorem 5.7, we may assume without loss of generality that $\mathcal{D}$ is fixed by all numerical multipliers. Suppose that a translate $g\mathcal{D}$ is also fixed by all numerical multipliers. Then

$$g\mathcal{D} = g^{t_i}\mathcal{D}^{(t_i)} = g^{t_i}\mathcal{D}$$

for $i=1,\ldots,s$. Thus, $g^{t_i-1}=1$, the identity of G. Also $g^v=1$. Since $(v,t_1-1,\ldots,t_s-1)=1$, there exist integers $b,a_1,\ldots,a_s$ such that $bv+a_1(t_1-1)+\ldots+a_s(t_s-1)=1$. Then

$$g=g^{bv+a_1(t_1-1)+\ldots+a_s(t_s-1)}=1.$$

Hence, $\mathcal{D}$ is the only translate fixed by all numerical multipliers.

Since the numerical multipliers lie in the center of the full multiplier group, it follows by Lemma 5.8 that $\mathcal{D}$ is the unique translate fixed by all multipliers.□

Theorem 5.10. Let $\mathcal{D}$ *be a* (v,k,λ)-*difference set in an abelian group* G. *Suppose that* $(v,k)=1$. *Then any group of multipliers fixes an equal number of points and blocks*.

In particular, some translate of $\mathcal{D}$ *is fixed by all multipliers*.

Proof. Let $f(g)$ be the product of the elements in the translate $g\mathcal{D}$. Then $f(g)=g^k f(1)$. Since $(v,k)=1$, this function ranges over all elements of G. Thus, a translate is completely specified by the product of its elements. After possibly replacing $\mathcal{D}$ by a translate, assume that $f(1)=1$.

Let Γ be a group of multipliers. Then Γ fixes the translate $g\mathcal{D}$ if and only if Γ fixes g^k. Thus Γ fixes an equal number of points and blocks. In particular, since Γ fixes the identity element, it also fixes some block.□

The theorems above are not only interesting from a theoretical poin of view. They can be used effectively to study a putative difference set. We consider, for example, n abelian difference set connected with a projective plane.

Proposition 5.11. Suppose that there exists a v,k,1)-difference set $\mathcal{D}$ in an abelian group G. Suppose that $_1, t_2, t_3, t_4$ are numerical multipliers of $\mathcal{D}$ and that

$$t_1 - t_2 = t_3 - t_4 \quad \text{(mod exponent of G).}$$

hen $t \equiv 0$ (mod exponent of G), where t is the least common ultiple of (t_1-t_2) and (t_1-t_3).

Proof. By Theorem 5.6 or Theorem 5.9, we may ssume that $\mathcal{D}$ is fixed by all numerical multipliers. Let us rite G additively. If $e \in \mathcal{D}$ then also t_1e, t_2e, t_3e, and t_4e re elements of $\mathcal{D}$. By hypothesis,

$$t_1e - t_2e = t_3e - t_4e.$$

ince every nonidentity element can be expressed in a unique ashion as a difference of elements of $\mathcal{D}$, we must have either $_1e = t_2e$ or $t_1e = t_3e$. That is, $(t_1-t_2)e = 0$ or $(t_1-t_3)e = 0$ where is the identity element of G. Thus for all elements e in $\mathcal{D}$,

$$te = 0,$$

here t is the least common multiple of (t_1-t_2) and (t_1-t_3). ence $t(e-f)=0$ for all $e, f, \in \mathcal{D}$. Since $(e-f)$ ranges over all lements of G, then necessarily

$$t \equiv 0 \quad \text{(mod exponent of G).}$$

his completes the proof.□

As an immediate consequence, an abelian (v,k,1)-ifference set in a group G cannot admit both 2 and 3 as ultipliers. (Take $t_1=3$, $t_2=2$, $t_3=2$ and $t_4=1$. Then the xponent of G must divide 1.) In particular if a projective lane of order n admits a regular abelian group then $6 \nmid n$. e mention some results in this spirit.

<u>Corollary 5.12</u>. <u>Suppose that there exists a projective plane of order</u> n <u>admitting a regular cyclic automorphism group</u> G. <u>Then</u> n <u>cannot be a multiple of</u> 6,10,14,15, 21,22,26,33,38,41,55,57,58,62 <u>or</u> 65.

<u>Proof</u>. Suppose, for example, that n is a multiple of 58. Then 2 and 29 are multipliers. Applying Proposition 5.11 with $t_1=32$, $t_2=29$, $t_3=4$ and $t_4=1$, we find that t=84. Thus the exponent of G divides 84. Since G is cyclic, its exponent is $v=n^2+n+1$. Since v is odd, v divides 21. However a projective plane of order divisible by 58 has far more than 21 points. Contradiction.

Table 5-3 lists the appropriate data.□

	multipliers	t_1	t_2	t_3	t_4	t	odd part of t
6	2,3	3	2	2	1	1	1
10	2,5	5	4	2	1	3	3
14	2,7	8	7	2	1	6	3
15	3,5	5	3	3	1	2	1
21	3,7	9	7	3	1	6	3
22	2,11	11	8	4	1	21	21
26	2,13	16	13	4	1	12	3
33	3,11	11	9	3	1	8	1
38	2,19	19	16	4	1	15	15
46	2,23	23	16	8	1	105	105
51	3,17	17	9	9	1	8	1
55	5,11	125	121	5	1	120	15
57	3,19	27	19	9	1	72	9
58	2,29	32	29	4	1	84	21
62	2,31	32	31	2	1	30	15
65	5,13	25	13	13	1	12	3

Table 5-3

We can improve upon Corollary 5.12-- replacing "cyclic" by "abelian"--if we exclude three of the possible divisors above.

Corollary 5.13. Suppose that there exists a projective plane of order n admitting a regular abelian automorphism group G. Then n cannot be a multiple of 6,10,14,15,
21,26,33,38,51,55,57,62 or 65.

Proof. Suppose that n is a multiple of one of these ntegers. By the proof of Corllary 5.12, the only possible rime factors of the exponent of G are 3 and 5. Hence

$$v=n^2+n+n+1 = 3^a5^b.$$

ow $n^2+n+1\equiv 0 \pmod 5$ and $n^2+n+1\equiv 0 \pmod 9$ have no solutions. ence b=0 and a=0 or 1. But then v=1 or 3, which is impos-ible.

Remark. If n were a multiple of 22,46 or 58, we ould have to consider also the possible prime factor 7. In his case we would want to investigate solutions to

$$v=n^2+n+1 = 7^c \quad \text{and} \quad v=n^2+n+1 = 3.7^c.$$

ere a simple congruence argument fails. Perhaps by con-idering the arithmetic of $Q(\sqrt{-3})$, a reader can determine all ntegral solutions to these equations.

For which integers $n\leq 100$ does there exist a pro-ective plane D admitting a regular abelian automorphism oup? There are 79 values of n satisfying the existence iteria of Chapter 2. Of these, 35 are prime powers, for ich appropriate planes certainly do exist. Corollary 5.13 les out all but 9 of the remaining 39 values. Of these, ur are multiples of 22,46 or 58; these can be excluded nce in these cases v does not have the form 7^c or 3.7^c. nally, the five remaining values--74,82,87,91 and 95 can be cluded by applying the First Multiplier Theorem. Hence,

Corollary 5.14. Suppose that there exists a pro-ctive plane of order n admitting a regular abelian auto-rphism group. If $n\leq 100$, then n is a prime power.

An analogous result to Proposition 5.11 holds for bitrary abelian (v,k,λ)-difference sets. It can be used similarly to prove nonexistence results about difference sets. For example, there does not exist an abelian $(v,k,2)$-difference set if n is a multiple of 12. (See Problem 6.)

Before closing this section we prove one result concerning blocks fixed by multipliers which applies even to nonabelian difference sets. The proof requires two deep theorems from group theory, which we mention without proof.

Theorem 5.15. (Schur-Zassenhaus) Let G be a group. Let A be a normal subgroup and let $B=G/A$. If the orders of A and B are relatively prime then A has a complement in G. That is, there exists a subgroup T such that $G=AT$ and $A \cap T=\{1\}$.

Moreover, if either A or B is solvable then all such complements are conjugate in G.

Theorem 5.16 (Feit-Thompson). A group of odd order is solvable.

Theorem 5.17. Let $\mathfrak{D}$ be a (v,k,λ)-difference set in a group G. If Γ is a group of multipliers of order relatively prime to v, then Γ fixes a translate of $\mathfrak{D}$.

Proof. Let $D=\text{dev}(\mathfrak{D})$. The group $H=G\Gamma$ acts as an automorphism group of D containing G as a normal subgroup. If the orders of G and Γ are relatively prime, then at least one has odd order. By the Feit-Thompson Theorem, at least one is solvable. By the Schur-Zassenhaus Theorem, any two complements of G in H are conjugate.

The subgroup H_p stabilizing a point p is a complemen for G in H. So is the subgroup H_B stabilizing a block B. Hence, they are conjugate. Now, Γ is just the subgroup stabilizing the identity element of G. Then $\Gamma = \alpha H_B \alpha^{-1}$ for some $\alpha \in H$. Thus, Γ fixes αB.□

We note in closing that there are analogous statements, about G/H-multipliers, to Theorems 5.7, 5.9, 5.10 and 5.17. The proofs carry over directly.

§5.4 FURTHER MULTIPLIER THEOREMS

The proof of Theorem 5.5 provides the point of departure for further multiplier theorems. Recall the outline of that proof. We let D be an arbitrary symmetric (v,k,λ)

sign with semi-regular automorphism group H, let π be a
rmutation of the points of D_H and let n_1 be a divisor of n.

(1) We defined the matrix S, showed that $SS^T=n^2I$ d $SJ=JS=nJ$, and showed that π is an automorphism of D_H if d only if $S=nP$ for some permutation matrix P.

(2) To show that $S=nP$, we saw that it was enough to ow that $S\equiv 0 \pmod{n_1}$, provided that either $n_1>h\lambda$ or $n_1=n$.

(3) To show that $S\equiv 0 \pmod{p^\alpha}$ for each prime power dividing n_1, we saw that it was sufficient that π induce automorphism of each module $M^H_{(\text{mod } p^\alpha)}$.

If we restrict our attention only to difference sets d potential multipliers (rather than allowing D and π to arbitrary) we can weaken the requirements on n_1 and π eded in steps (2) and (3) above. In this section we work step (3); in the next section, on step (2). Our principal ol is the following result:

Lemma 5.18. Suppose that D is the development of a fference set in a group G, that H is a normal subgroup of G d that π is an automorphism of G/H.

Then (after possibly rearranging its rows) S is a H-matrix.

Proof. The matrix A_H is a G/H-matrix. Check that, ter possibly rearranging its rows, $(A_H)^\pi$ is a G/H-matrix. nce the same is true about $S=(A_H)^\pi A_H - h\lambda J$.□

Since it does not matter in steps (1),(2) and (3) we rearrange the rows of S, we may assume without harm at S is a G/H-matrix. We can use this information to prove e following theorem.

Theorem 5.19. Let D be a (v,k,λ)-difference set in abelian group G and let H be a subgroup of order h and dex w. Let M^H be the $\mathbb{Z}$-module spanned by the rows of the ntracted incidence matrix A_H.

Let n_1 be a divisor of n with canonical prime fac- rization $n_1=p_1^{\alpha_1} \dots p_s^{\alpha_s}$.

Suppose that γ is an automorphism of G/H.

If (1) *for* $i=1,\ldots,s$, *either*

(a) γ *induces an automorphism of* $M^H_{(\text{mod } p_i^{\alpha_i})}$, *or*,

(b) p_i *is semiprimitive (mod exponent of* G/H) *and*,

(2) $n_1>h\lambda$ *or* $n_1=n$,

then γ *defines an automorphism of* D_H.

Proof. The proof proceeds as in steps (1) and (2). Now consider the prime p_i. If γ induces an automorphism of $M^H_{(\text{mod } p_i^{\alpha_i})}$ then $S\equiv 0 \pmod{p_i^{\alpha_i}}$, as before. Suppose, on the other hand, that p_i is semiprimitive (mod exponent of G/H). Since $SS^T\equiv 0 \pmod{p_i^{2\alpha_i}}$ then $S\equiv 0 \pmod{p_i^{\alpha_i}}$ by Proposition 4.16. In either case $S\equiv 0 \pmod{p_i^{\alpha_i}}$ for $i=1,\ldots,s$.□

We mention two consequences.

Theorem 5.20. *Let* $\mathbb{D}$ *be a* (v,k,λ)-*difference set in an abelian group* G. *Let* n_1 *be a divisor of* n *and let* $n_1=p_1^{\alpha_1}\ldots p_s^{\alpha_s}$ *be its canonical prime factorization*.

Suppose that H *is a subgroup of order* h *and index* w. *Suppose that* t *is an integer relatively prime to* w.

If (1) *for* $i=1,\ldots,s$, *there exists an integer* $j=j(i)$ *such that either*

(a) $p_i^j\equiv t$ (mod exponent of G/H) *or*

(b) $p_i^j\equiv -1$ (mod exponent of G/H), *and*

(2) $n_1>h\lambda$ *or* $n_1=n$,

then t *is a* G/H-*multiplier*.

Proof. Theorem 5.19 and Fact 2.□

Suppose, for example, that there exists a (221,45,9) difference set in $\mathbb{Z}_{221}$. Let H be the subgroup of order 13, let $n_1=n=36$ and let $t=3$. Since $2^8\equiv -1 \pmod{17}$ and $3^1\equiv t \pmod{17}$, then 3 is a multiplier. Now "multiplication by 3" has orbits of size 1 and 16 on $\mathbb{Z}_{17}$. Thus a fixed row of A_H has the form

$$\underbrace{(a,b,\dots,b)}_{16\text{ times}}$$

up to order, where a and b must satisfy

$$a^2 + 16b^2 = 36 + 13\cdot 9 = 153$$

$$a^2 + 16b = 45$$

These equations have no nonnegative integral solution. Hence no such difference set exists. (Note that Theorem 5.5 alone would not show that 3 must be a multiplier.)

We also obtain our first theorem which applies even to non-numerical multipliers.

Theorem 5.21. Let $\mathcal{D}$ *be a* (v,k,λ)*-difference set in an abelian group* G. *Let* n_1 *be a divisor of* n. *Let* H *be a subgroup of* G *of order* h.

If (1) n_1 *is semiprimitive* (mod exponent of G/H), *and*

(2) $n_1 > h\lambda$ *or* $n_1 = n$,

then every automorphism of G/H *is a* G/H-*multiplier*.

Proof. This is the special case of Theorem 5.19 in which every prime dividing n_1 satisfies condition (1)(b).□

The theorem rules out, for example a (352,27,2)-difference set in six of the seven abelian groups of order 352. Let $\mathcal{D}$ be such a difference set in an abelian group G.

(1) Suppose that $G = \mathbb{Z}_{11} \times (\mathbb{Z}_2)^5$. Let H be the subgroup of order 11. Then $(G/H) = (\mathbb{Z}_2)^5$. Take $n_1 = 25$ in Theorem 5.21 and conclude that every automorphism of G/H is a G/H-multiplier. The full automorphism group of G/H is GL(5,2) and is transitive on nonidentity elements. By Theorem 5.10, it fixes a block B of $\mathcal{D}_H$. Now a fixed row must be of the form

$$\underbrace{(a,b,\dots,b)}_{31}$$

and a and b must satisfy

$$a^2 + 31b^2 = 25 + 11.2 = 47$$

and

$$a + 31b = 27.$$

These equations have no nonnegative integral solution, whence no such difference set exists.

(2) Suppose that $G = \mathbb{Z}_{11} \times \mathbb{Z}_4 \times (\mathbb{Z}_2)^3$. Let H be a subgroup of order 22 such that $(G/H) \simeq (\mathbb{Z}_2)^4$. A similar argument shows that

$$a^2 + 15b^2 = 69$$

and

$$a + 15b = 27$$

must have a nonnegative integral solution in common, which is false.

(3) The four remaining noncyclic groups require somewhat more work. Suppose that $G = \mathbb{Z}_{11} \times \mathbb{Z}_8 \times (\mathbb{Z}_2)^2$, or $\mathbb{Z}_{11} \times (\mathbb{Z}_4)^2 \times \mathbb{Z}_2$, or $\mathbb{Z}_{11} \times \mathbb{Z}_8 \times \mathbb{Z}_4$, or $\mathbb{Z}_{11} \times \mathbb{Z}_{16} \times \mathbb{Z}_2$. Let $K \subseteq H \subseteq G$ be subgroups such that $G/K = \mathbb{Z}_4 \times \mathbb{Z}_2$ and $G/H = \mathbb{Z}_2 \times \mathbb{Z}_2$. Arguing as before leads to the equations

$$a^2 + 3b^2 = 201$$

and

$$a + 3b = 27 ,$$

which have the unique nonnegative integral solution $(a,b) = (3,8)$. Thus a row of A_H must have the form (3,8,8,8), up to order. Consider A_K. By Theorem 5.6, 5 is a G/K-multiplier. Consider a row of A_K fixed by this multiplier, which when contracted gives (3,8,8,8). It must have the form

$$(u,4,v,4,3-u,4,8-v,4)$$

where the coordinates correspond to the elements (0,0),(1,0),((3,0),(0,1),(1,1),(2,1),(3,1) in $\mathbb{Z}_4 \times \mathbb{Z}_2$, respectively.

nce

$$A_K A_K^T = 25\ I + 88\ J$$

$$A_K J = 27\ J$$

e sum of the squares of the entries in the row must be 103. nce $(u,v) = (1,3),(1,5),(2,3)$ or $(2,5)$. However, with any these values, the product of a row of A_K with its translate the group element $(0,1)$ is 90 or 86 rather than 88. Hence appropriate matrix A_K exists and no such difference set ists.

.5 STILL FURTHER MULTIPLIER THEOREMS

We now turn to weakening the requirement on n_1 in ep (2) of the outline given in the previous section. Suppse that $S \equiv 0 \pmod{n_1}$ but that $n_1 \nmid h\lambda$ and $n_1 \neq n$. We can use e fact that S is a G/H-matrix to show that, under appropriate potheses, S must nevertheless be a multiple of a permutation trix.

Set $T = \frac{1}{n_1} S$. Then T is an integral G/H-matrix ch that

$$TT^T = (n/n_1)^2 I$$

$$\text{and} \quad TJ = (n/n_1)J.$$

e next theorem shows that T (and therefore S) is a multiple a permutation matrix provided only that $|G/H|$ is relatively ime to a certain integer depending only n/n_1.

Theorem 5.22. For every positive integer m, there ists an integer M(m) such that if K is a finite abelian oup with order w relatively prime to M(m) then the only matrices satisfying the equations

$$UU^T = m^2 I$$

$$\text{and} \quad UJ = mJ$$

are multiples of permutation matrices.

We can define $M(m)$ as follows: $M(1)=1$. For $m>1$, let $M(m)$ be the product of the distinct prime factors of

$$m,\ M(m^2/p^{2e}),\ p-1,\ p^2-1,\ldots p^{u(m)}-1$$

where p is a prime dividing m such that $p^e \| m$ and where $u(2)=3$, $u(3)=5$, $u(4)=7$ and $u(m)=\frac{1}{2}(m^2-m)$ for $m \geq 5$.

Remark. Note that $M(m)$ is defined inductively for each positive integer m, but not uniquely so. For uniqueness we could require, for example, that p be the largest prime factor of m. However, the theorem is true no matter how p is chosen in the inductive calculation of $M(m)$.

Proof. Clearly $M(1)=1$. Let $m>1$ have exactly $s>0$ distinct prime factors. Assume that the theorem is true for each integer with fewer than s distinct prime factors. Let p be a prime factor of m and suppose that $p^e \| m$. Then, by inductive hypothesis, $M(m^2/p^{2e})$ is defined and satisfies the theorem. Define $M(m)$ as above. Let K be an abelian group of order w relatively prime to $M(m)$.

Step 1. Let U be a K-matrix satisfying the equation

$$UU^T = m^2 I \qquad (*)$$

and

$$UJ = mJ. \qquad (**)$$

Consider the $\hat{\mathbb{Z}}_p$-module M_p generated by the rows of U. The automorphism α of K given by $\alpha(g)=g^p$ for $g \in K$ preserves M_p, by Fact 2. Thus the rows of U^α are elements of M_p. Write $U^\alpha = UQ$, for an appropriate permutation matrix Q.

From (*), the dot product of elements of M_p is a multiple of p^{2e}. Hence

$$U^\alpha U^T \equiv 0 \pmod{p^{2e}}.$$

Step 2. Let $V = p^{-2e} U^\alpha U^T$. Then V is a G-matrix such that

$$VV^T = (p^{-2e})^2 U^\alpha U^T U (U^\alpha)^T = (m^2/p^{2e})^2 I$$

and $$VJ = p^{-2e} U^\alpha U^T J = (m^2/p^{2e}) J.$$

.nce w is relatively prime to M(m), then w is also relatively ·ime to $M(m^2/p^{2e})$. By inductive hypothesis, V must be a ıltiple of a permutation matrix. Say $V=(m^2/p^{2e})P$. Then

$$U^\alpha U^T = m^2 P.$$

Step 3. Since $UU^T = m^2 I$ and $U^\alpha U^T = m^2 P$, then

$$UQ = U^\alpha = PU.$$

·guing as in the proof of Theorem 3.1, we conclude that since ıe permutation associated with the matrix Q has a fixed point,) does the permutation associated with the matrix P. Since ' = PU, at least one row of U is then left fixed when its ›ordinates are permuted according to α.

Step 4. Let $(a_{g_1}, \ldots, a_{g_w})$ be such a fixed row, ıere for convenience we let g_1 be the identity element of K. : study the action of α on K. Now $\alpha^f(g) = g^{p^f}$. Since w is ːlatively prime to p, p-1, $p^2-1, \ldots, p^{u(m)}-1$ by hypothesis, ıe u(m)+1 elements

$$g, g^p, \ldots, g^{p^{u(m)}}$$

·e distinct provided that g is not the identity. Hence every ·bit of α has size at least u(m)+1, except for the singelton ·bit $\{g_1\}$. Since $(a_{g_1}, \ldots, a_{g_w})$ is fixed under α, any value ıat occurs once as one of the a_{g_i} with i>1 must occur in ıis way at least u(m)+1 times. We shall see that this ıplies that $a_{g_i}=0$ for $i \neq 1$.

Suppose that $a_{g_i}=k$ for some index i>1. Then

$$m^2 = \sum_{i=1}^{w} (a_{g_i})^2 \geq k^2(u(m)+1).$$

Thus, for $m \geq 2$, we have

$$4 \geq \frac{m^2}{\frac{1}{2}(m^2-m)} \geq \frac{m^2}{u(m)+1} \geq k^2.$$

Hence $k=-1,0$ or 1.

Suppose that $a_{g_1}=a$, that $a_{g_i}=+1$ for π values of $i>1$ and that $a_{g_i}=-1$ for η values of $i>1$. From (*) and (**), we have

$$a^2 + \pi + \eta = m^2$$

and

$$a + \pi - \eta = m.$$

Thus

$$\eta = \tfrac{1}{2}(m^2-m) - \tfrac{1}{2}(a^2-a) \leq u(m).$$

The value -1 occurs too few times, unless $\eta=0$. In this case,

$$\pi = 2m-1 \leq u(m).$$

The value $+1$ occurs too few times, unless $\pi=0$. This shows that the fixed row in question must be $(m,0,\ldots,0)$. Hence U is m times a permutation matrix, completing the proof.□

According to the theorem, the first few values of M(m) are

$M(2) = 2\cdot3\cdot7$

$M(3) = 2\cdot3\cdot5\cdot11\cdot13$

$M(4) = 2\cdot3\cdot5\cdot7\cdot31\cdot127$

$M(5) = 2\cdot3\cdot5\cdot7\cdot11\cdot13\cdot19\cdot31\cdot71\cdot313\cdot521\cdot829\cdot19531$

$M(6) = 2\cdot3\cdot5\cdot7\cdot11\cdot13\cdot23\cdot41\cdot61\cdot73\cdot547\cdot1093\cdot3851$
$4561\cdot797161.$

These are, in no sense, best possible values. By arguing more carefully in Step 4 we can make _ad hoc_ improvements upon individual values of M(m).

Proposition 5.23. The previous theorem remains
ue with

(1) $M(2) = 2\cdot 7$,

(2) $M(3) = 2\cdot 3\cdot 11\cdot 13$,

(3) $M(4) = 2\cdot 3\cdot 7\cdot 31$.

Proof. (1) Take p=2 and proceed with the proof til Step 4. When we assumed in the proof that $(w,2\cdot 3\cdot 7)=1$, concluded that any orbit other than $\{g_1\}$ had size at least 2)+1=4. If we merely assume that $(w,2\cdot 7)=1$, then K may ssibly have elements of order 3. If g is such an element, en $\{g,g^2\}$ will be an orbit of α of size 2. In other words, must now consider the additional possibility that there e orbits of size 2.

Since $\sum(a_{g_i})^2=4$, there can be at most two orbits size 2 corresponding to nonzero entries. Say $a_x=0$ except ssibly for $x \in \{g_1\} \cup \{g,g^2\} \cup \{h,h^2\}$, where g and h are ements of order 3. Then $a_g=a_{g^2}$ and $a_h=a_{h^2}$. Therefore,

$$(a_{g_1})^2 + 2(a_g)^2 + 2(a_h)^2 = 4$$

and

$$(a_{g_1}) + 2(a_g) + 2(a_h) = 2.$$

inspection the only solution is $(a_{g_1},a_g,a_h) = (2,0,0)$. nce, even if we allow orbits of size 2, the only solutions $UU^T=4I$ and $UJ=2J$ in K-matrices are multiples of permutation trices.

(2) Argue as in part (1). Take p=3. We must take to account the possibility that K has elements of order 5, ich would lie in an orbit of α of size 4 (since $3^4\equiv 1$ od 5)). Again, there can be at most two such orbits with nzero entries. However

$$(a_{g_1})^2 + 4(a_g)^2 + 4(a_h)^2 = 9$$

$$(a_{g_1}) + 4(a_g) + 4(a_h) = 3$$

s only the "trivial" solution (3,0,0).

(3) Since $2^4 \equiv 1 \pmod 5$ and $2^7 \equiv 1 \pmod{127}$ we must consider orbits of size 4 and 7. Studying the relevant equations we find one possible solution apart from the "trivial" solution. We might have

$$a_x = \begin{cases} -3 & \text{if } x = g_1 \\ +1 & \text{if } x \in \{g, g^2, g^4, g^8, g^{16}, g^{32}, g^{64}\} \\ 0 & \text{otherwise} \end{cases}$$

where g is a fixed element of order 127. However, such a row cannot occur in the matrix U since the dot product of this row with its translation by g is -2. This dot product must be 0 since $UU^T = 9I$.□

Improvements can be made similarly in other values of M(m). (For example, the factor 797161 is not necessary in M(6).) These arguments, however, are completely *ad hoc*.

In view of the discussion at the beginning of this section, we have the following result.

Theorem 5.24. Theorems 5.6, 5.19, 5.20 and 5.21 remain true if we replace the condition

$$"n_1 > h\lambda \quad \text{or} \quad n_1 = n"$$

by the condition

$$"n_1 > h\lambda \quad \text{or} \quad (w, M(n/n_1)) = 1".$$

We mention two applications:

(1) Suppose that there exists a (199,45,10)-differenc set in an abelian group G. Apply Theorem 5.6, extended as above, with $n_1=7$, $t=7$ and $h=1$. Since $(199, M(5))=1$, then 7 is a multiplier. However, "multiplication by 7" has orbits of sizes 1, 99 and 99. Contradiction.

(2) We can show that any (263,131,65)-difference set is equivalent to the difference set associated with H(263). Applying Theorem 5.6, extended as above, we find that t=11 must be a multiplier. (Let $n_1=11$ and $h=1$.) The order of 11(mod 263) is 131. So, if B is a translate fixed by this multiplier and if $a \in B$ is a nonzero element, then $B = \{a, 11a, \ldots, 11^{130}a\}$. Hence, any two such difference sets are equivalent.

PROBLEMS - CHAPTER 5

The first five problems concern cyclic projective anes--that is, cyclic $(n^2+n+1, n+1, 1)$-difference sets.

Show that the number of blocks fixed by all multipliers either 1 or 3, with the latter case possible only if $n \equiv 1$ od 3). (Hint: consider the multiplier n.)

Let t be a multiplier. Show that t fixes m^2+m+1 points r some integer $m \geq 0$. If $m \geq 2$, the fixed points form a cyclic ojective plane.

Show that 2 is a multiplier if and only if n is even.

Show that 3 is a multiplier if and only if n is a multiple 3.

Show that there do not exist cyclic projective planes of der n= 74, 82, 87, 91 or 95.

Let D be an abelian (v,k,2)-difference set. Show that 0 (mod 12).

Prove Corollary 4.42. (Hint: Show that all divisors of have multiplicative order q(mod p). Hence show that all adratic residues (mod p) are multipliers of the difference t. Suppose that n is prime. By considering the factoriza- on of n in $Q(\sqrt{-p})$, observe that $x^2+py^2=4n$ has at most one lution up to choice of signs. By considering the equations Theorem 4.41 show that $n=\frac{1}{4}(p^m+1)$ with m necessarily odd. wever in this case n is not prime.)

Let U be the group of units modulo p^k, where p is an odd ime.

(i) Show that $|U| = p^{k-1}(p-1)$.

(ii) Show that F_p^* is a cyclic group of order p-1 and hence that U contains elements of order p-1. (Hint: If d is the exponent of the group--that is, the least common multiple of orders of the elements--then $x^d-1=0$ has p-1 solutions in F_p.)

(iii) Let U_1 be the subgroup of U consisting of all elements congruent to 1 (mod p). Show that U_1 is a cyclic group of order p^{k-1}. (Hint: Let $\alpha=1+p$. By induction show that $\alpha^{p^i}\equiv 1 \pmod{p^{i+1}}$ but $\alpha^{p^i}\not\equiv 1 \pmod{p^{i+2}}$.)

(iv) Observe that U is cyclic.

9. Let U be the group of units modulo 2^k with $k\geq 3$.

(i) Let U_2 be the subgroup of U consisting of all elements congruent to 1 (mod 4). Show that U_2 is a cyclic group of order 2^{k-2}. (Hint: Let $\alpha=1+2^2$.)

(ii) Show that U is a direct product of a cyclic group of order two and a cyclic group of order 2^{k-2}. Explicitly identify the three elements of order two.

The following problems ask you to use multiplier techniques to show the nonexistence of certain difference sets. The multiplier information should allow you to eliminate the difference set immediately or to narrow it down to a few candidates which can be eliminated by direct hand or machine calculation, if you have time. (In any case, no problem requires more than three possible candidates.)

10. Show that there does not exist a cyclic (81,16,3)-difference set.

11. Show that there does not exist a (81,16,3)-difference set in $\mathbb{Z}_3 \times \mathbb{Z}_{27}$.

12. Show that there does not exist a cyclic (85,21,16)-difference set.

13. Show that there does not exist a cyclic (253,28,3)-difference set.

. Show that there does not exist a cyclic (103,34,11)-
fference set.

. Show that there does not exist a (121,40,13)-difference
t in $\mathbb{Z}_{11} \times \mathbb{Z}_{11}$.

. Show that there does not exist a cyclic (415,46,5)-
fference set.

. Show that there does not exist a cyclic (491,50,5)-
fference set.

. Show that there does not exist a cyclic (70,24,8)
fference set.

e any methods you please--including educated trial and error
to answer the last four problems.

. Show that there does not exist a cyclic (352,27,2)-
fference set.

. Show that there does not exist a cyclic (375,34,3)-
fference set.

. Show that there does not exist a cyclic (165,41,10)-
fference set.

. Show that there does not exist a cyclic (231,46,9)-
fference set.

NOTES TO CHAPTER 5

§5.1 The First Multiplier Theorem was proven originally for cyclic groups by Hall [48], who used quite different methods. It was later extended to abelian groups and to the more general Second Multiplier Theorem. See [53], [19]. The present interpretation, involving a transfer from "local" to "global" automorphisms, was proven by Lander [79].

§5.3 McFarland and Rice [99] proved Theorems 5.7 and 5.9. Proposition 5.11 and Corollary 5.12 were proved by Hall [48]. Feit and Thompson's classic paper is [39]. For a proof of the Schur-Zassenhaus result, see e.g. [114]. Theorem 5.17 was pointed out to me by Cameron.

§5.4 The results of this section are due to Lander [79].

§5.5. The results of this section are taken from McFarland's thesis [96].

Problems. Various results like those in Problems 1-4 are collected in Baumert's survey book [12].

6. OPEN QUESTIONS

In this chapter we discuss some important open
estions and conjectures concerning difference sets.
swering these questions apparently will require new and
werful techniques or clever counterexamples; perhaps the
ader will try to wrestle with one or more of these
estions.

.1 EXISTENCE

The broadest open question is of course: For what ,k,λ) and abelian groups G, does there exist a (v,k,λ)-fference set in G? Despite many examples and many non-istence theorems, there is no complete conjectured answer. partial guess, related to Conjecture 2.4 is the following.

Conjecture 6.1. For each $\lambda>1$, there exists (up to omorphism) only finitely many abelian (v,k,λ)-difference ts.

There is slightly more evidence amassed in favor of njecture 6.1 than Conjecture 2.4. Specifically, Dickey d Hughes [64] have shown that the only abelian (v,k,2)-fference sets with $k \leq 5000$ are the known ones with 3,4,5,6 or 9.

Table 6-1, at the end of this chapter, lists the ate of my knowledge concerning abelian difference sets with 50. There are 268 quadruples (v,k,λ,G) where G is an elian group of order v and (v,k,λ) satisfy $(v-1)\lambda = k(k-1)$, hutzenberger's Theorem, the Bruck-Ryser-Chowla conditions d $k \leq v/2$. Of these, 65 correspond to known difference sets, 8 are shown not to exist and 25 are undecided. I would be ateful to hear from readers who can complete any of the ssing entries in the table.

§6.2 CYCLIC SYLOW SUBGROUPS

No one has yet discovered an example of a cyclic (v,k,λ)-difference set in which $(v,n)\neq 1$. This circumstance seems to be special to cyclic groups, since many constructions are known of non-cyclic difference sets in which v and n share a factor. Ryser [115] has conjectured:

Conjecture 6.2. *If* $\mathcal{D}$ *is a* (v,k,λ)*-difference set in a cyclic group* G, *then* $(v,n)=1$.

Baumert [12] has enumerated all parameters (v,k,λ) with $k\leq 100$ for which cyclic difference sets exist and in no case is $(v,n)>1$. Despite this evidence, no real progress has been made in settling the conjecture, or even in pinpointing just what property of cyclic groups "obstructs" such a difference set. While I also have no idea how to prove this conjecture, I might suggest that the special properties of the group ring $\mathbb{F}_pG$, when G is a cyclic p-group, might help shed some light upon the issue. Based on the work in §4.6, I am led to the following conjecture.

Conjecture 6.3. *If* $\mathcal{D}$ *is a* (v,k,λ)*-difference set in an abelian group* G *with a cyclic Sylow* p*-subgroup, then* p *does not divide* (v,n).

This conjecture clearly implies Ryser's. In support of Conjecture 6.2 I offer the following evidence.

(1) The conjecture is true whenever p is self-conjugate (modulo exponent of G). This is of course simply Theorem 4.30.

(2) The conjecture is true for all abelian difference sets with $k\leq 75$. We may check this for $k\leq 50$ by consulting the list in Table 6-1. Each of the 34 cases to be considered has been marked with an asterisk; nonexistence is shown in all cases. For $51\leq k\leq 75$, there are an additional 22 cases to consider. All but two are disposed of by results in Chapter 4. That the remaining two cannot exist follows from a calculation of Rumsey, reported in Baumert [12]. The cases are presented in Table 6-2, at the end of this chapter.

5.3. CYCLIC PROJECTIVE PLANES

The fact that the only projective planes known to dmit a cyclic automorphism group are the planes PG(2,q) has ed to the following conjecture.

Conjecture 6.4. Let D be a projective plane of rder n admitting a cyclic regular automorphism group. Then must be a prime power and D must be isomorphic to PG(2,n).

There is a great deal of evidence supporting the elief than n must be a prime power. For example, the statement is true for all $n<74$ by virtue of Corollary 5.12. sing similar, although more sophisticated tests, Evans and ann [37] extended this to $n\le 1600$. According to Dembowski 33, p. 209], Keiser has extended the bound still further o $n\le 3600$ in an unpublished work.

Only one major result has been proven in the eneral direction of understanding the structure of a cyclic rojective plane. Ott [107] proved the following deep esult.

Theorem 6.5. Let D be a projective plane of order admitting a cyclic regular automorphism group G. Let Γ e the full automorphism group of D. Either

(1) G is a normal subgroup of Γ (in which case Γ is generated by G together with its multipliers); or

(2) D is isomorphic to PG(2,n).

n immediate corollary of Ott's result is the following.

Corollary 6.6. Let D be a projective plane of rder n, admitting two distinct cyclic regular automorphism roups. Then D is isomorphic to PG(2,n).

Since the only known cyclic projective planes are he PG(2,q), it is easy to make up reasonable conjectures bout cyclic projective planes. Simply conjecture that any roperty enjoyed by the PG(2,q) must hold for all cyclic rojective planes.

Since no further abelian (v,k,1)-difference sets re known, we could ask more generally whether all abelian rojective planes must be isomorphic to some PG(2,q).

§6.4. MULTIPLIER THEOREMS

The condition that $p>\lambda$ is crucial to all known proofs of the First Multiplier Theorem. However, no examples are known showing that this restriction is necessary. The following has been conjectured.

Conjecture 6.7. Let $\mathcal{D}$ be a (v,k,λ)-difference set in an abelian group G. Let p be a prime which divides n but does not divide v.

Then p is a multiplier of $\mathcal{D}$.

Virtually all further multiplier theorems have arisen in an attempt to weaken the condition $p>\lambda$. However, it seems quite difficult to dispose of it entirely.

We might also mention that very little is known about multipliers of nonabelian difference sets. In fact, the only known result does not even concern multipliers directly. Let $\mathcal{D}$ be a (v,k,λ)-difference set in a group G. We say that an integer t relatively prime to v is a weak multiplier if the set

$$\mathcal{D}^{(t)} = \{x^t \mid x \in \mathcal{D}\}$$

is a left translate $g\mathcal{D}$ of $\mathcal{D}$. Also, say that D is central if

$$g\mathcal{D}g^{-1} = \mathcal{D}$$

for all $g \in G$. We have the following generalization of the Second Multiplier Theorem.

Theorem 6.8. Let $\mathcal{D}$ be a central (v,k,λ)-difference set in a group G. Let n_1 be a divisor of n and let $n_1 = p_1^{\alpha_1} \cdots p_s^{\alpha_s}$ be its canonical prime factorization.

Suppose that t is an integer relatively prime to v.

If: (1) for $i=1,\ldots,s$, there exists an integer $j=j(i)$ such that $t=p_i^j$ (mod exponent of G), and

(2) $n_1 > \lambda$

then t is a weak multiplier of $\mathcal{D}$.

The proof is virtually identical to that of Theorem .4. Only one change is required. Instead of Fact 2, we equire the fact (which we do not prove here, but which is not ifficult if the reader knows the representation theory of onabelian groups in the semisimple case) that p is a "weak ultiplier" of every two-sided ideal in $\hat{\mathbb{Z}}_pG$. The condition hat D is central assures us that all modules under considera- ion are two-sided ideals.

I do not know whether the condition that D be central ay be dropped entirely from the statement of the theorem; ertainly it is crucial to the proof. (Ott [108] states heorem 6.8 without the condition that D is central. However, is proof, which is by quite different methods, in fact equires D to be central.)

6.5 TABLES

We close this monograph with the two tables mentioned n §§6.1-6.2. Table 6-1 presents the state of my knowledge oncerning abelian (v,k,λ)-difference sets with k≤50.

ABLE 6-1

	(v,k,λ)	n	GROUP	EXIST?	EXAMPLE OR NONEXISTENCE PROOF
.	(7,3,1)	2	(7)	YES	PG(2,2)
.	(13,4,1)	3	(13)	YES	PG(2,3)
.	(21,5,1)	4	(3)(7)	YES	PG(2,4)
.	(11,5,2)	3	(11)	YES	H(11)
.	(31,6,1)	5	(31)	YES	PG(2,5)
.	(16,6,2)	4	(16)	NO	THEOREM 4.30 or 4.31

	(v,k,λ)	n	GROUP	EXIST ?	EXAMPLE OR NONEXISTENCE PROOF
7.			(8)(2)	YES	Ch.4, Example $4\frac{1}{2}$
8.			$(4)^2$	YES	Ch.4, Example 4
9.			$(4)(2)^2$	YES	Ch.4, Example 4
10.			$(2)^4$	YES	Ch.4, Example 4
11.	(15,7,3)	4	(3)(5)	YES	PG(3,2)
12.	(57,8,1)	7	(3)(19)	YES	PG(2,7)
13.	(73,9,1)	8	(73)	YES	PG(2,8)
14.	(37,9,2)	7	(37)	YES	Ch.7, Example 7
15.	(25,9,3)	6	(25)	NO	THEOREM 4.4
16.			$(5)^2$	NO	THEOREM 4.4
17.	(19,9,4)	5	(19)	YES	H(19)
18.	(91,10,1)	9	(7)(13)	YES	PG(2,9)
19.	(31,10,3)	7	(31)	NO	THEOREM 5.3
20.	(111,11,1)	10	(3)(37)	NO	THEOREM 4.4
21.	(56,11,2)	9	(8)(7)	NO	PROPOSITION 4.18, 4.19 or 4.20
22.			(2)(4)(7)	NO	THEOREM 4.5
23.			$(2)^3(7)$	NO	THEOREM 4.5
24.	(23,11,5)	6	(23)	YES	H(23)

(v,k,λ)	n	GROUP	EXIST ?	EXAMPLE OR NONEXISTENCE PROOF
(133,12,1)	11	(7)(19)	YES	PG(2,11)
(45,12,3)	9	(9)(5)	NO	THEOREM 4.30
		$(3)^2(5)$	YES	Ch.4, Example 4
(157,13,1)	12	(157)	NO	THEOREM 4.4
(79,13,2)	11	(79)	NO	THEOREM 5.3
(40,13,4)	9	(8)(5)	YES	PG(3,4)
		(2)(4)(5)	NO	PROPOSITION 4.18, 4.19 or 4.20
		$(2)^3(5)$	NO	THEOREM 4.5
(27,14,6)	7	(27)	NO	THEOREM 4.38
		(3)(9)		
		$(3)^2$	YES	H(27)
(183,14,1)	13	(3)(61)	YES	PG(2,13)
(71,15,3)	12	(71)	NO	THEOREM 5.4
(36,15,6)	9	(4)(9)	NO	THEOREM 4.30
		$(2)^2(9)$	NO	THEOREM 4.30
		$(4)(3)^2$	YES	Ch. 4, Problem 6
		$(2)^2(3)^2$	YES	Ch.4, Example 3
(31,15,7)	8	(31)	YES	H(31) and PG(4,2)

	(v,k, λ)	n	GROUP	EXIST?	EXAMPLE OR NONEXISTENCE PROOF
43.	(241,16,1)	15	(241)	NO	THEOREM 4.4
44.	(121,16,2)	14	(121)	NO	THEOREM 4.4
45.			$(11)^2$	NO	THEOREM 4.4
46.	(81,16,3)	13	(81)	NO	Ch.5, Problem 10
47.			(3)(27)	NO	Ch.5, Problem 11
48.			$(9)^2$		
49.			$(3)^2(9)$		
50.			$(3)^4$	NO	Ch.4, Problem 22
51.	(61,16,4)	12	(61)	NO	THEOREM 4.4
52.	(49,16,5)	11	(49)	NO	THEOREM 5.3
53.			$(7)^2$	NO	THEOREM 4.41
54.	(41,16,6)	10	(41)	NO	THEOREM 4.4
55.	(273,17,1)	16	(3)(7)(13)	YES	PG(2,16)
56.	(69,17,4)	13	(3)(23)	NO	THEOREM 5.3
57.	(35,17,8)	9	(5)(7)	YES	Ch.4, Example 8
58.	(307,18,1)	17	(307)	YES	PG(2,17)
* 59.	(154,18,2)	16	(2)(7)(11)	NO	PROPOSITION 4.19
60.	(343,19,1)	18	(343)	NO	THEOREM 4.5

	(v,k, λ)	n	GROUP	EXIST ?	EXAMPLE OR NONEXISTENCE PROOF
1.			(7)(49)	NO	THEOREM 4.5
2.			$(7)^3$	NO	THEOREM 4.5
3.	(115,19,3)	16	(5)(23)	NO	THEOREM 4.39
4.	(39,19,9)	10	(3)(13)	NO	THEOREM 4.4
5.	(381,20,1)	19	(3)(127)	YES	PG(2,19)
6.	(191,20,2)	18	(191)	NO	THEOREM 5.3
7.	(96,20,4)	16	(32)(3)	NO	THEOREM 4.30
8.			(2)(16)(3)	NO	THEOREM 4.32
9.			(4)(8)(3)		
0.			$(2)^2(8)(3)$		
1.			$(2)(4)^2(3)$		
2.			$(2)^3(4)(3)$	YES	Ch.4, Example 4½
3.			$(2)^5(3)$	YES	Ch.4, Example 4
4.	(421,21,1)	20	(421)	NO	THEOREM 4.4
5.	(211,21,2)	19	(211)	NO	THEOREM 5.3
6.	(85,21,5)	16	(5)(17)	NO	Ch. 5, Problem 12
7.	(71,21,6)	15	(71)	NO	THEOREM 5.4

	(v,k,λ)	n	GROUP	EXIST ?	EXAMPLE OR NONEXISTENCE PROOF
78.	(43,21,10)	11	(43)	YES	H(43)
79.	(155,22,3)	19	(5)(31)	NO	THEOREM 4.4
* 80.	(78,22,6)	16	(2)(3)(13)	NO	PROPOSITION 4.19 OR THEOREM 4.33
81.	(47,23,11)	12	(47)	YES	H(47)
82.	(553,24,1)	23	(7)(79)	YES	PG(2,23)
83.	(139,24,4)	20	(139)	NO	THEOREM 4.5
* 84.	(70,24,8)	16	(2)(35)	NO	Ch.5, Problem 18
85.	(601,25,1)	24	(601)	NO	COROLLARY 5.12
86.	(301,25,2)	23	(7)(43)	NO	THEOREM 5.3
87.	(201,25,3)	22	(3)(67)	NO	THEOREM 4.4
88.	(121,25,5)	20	(121)	NO	THEOREM 4.5
89.			$(11)^2$	NO	THEOREM 4.5
90.	(101,25,6)	19	(101)	YES	Ch.4, Example 7
91.	(61,25,10)	15	(61)	NO	THEOREM 4.4
92.	(51,25,12)	13	(3)(17)	NO	THEOREM 4.4
93.	(651,26,1)	25	(3)(7)(31)	YES	PG(2,25)
94.	(131,26,5)	21	(131)	NO	THEOREM 5.3

	(v,k,λ)	n	GROUP	EXIST?	EXAMPLE OR NONEXISTENCE PROOF
5.	(66,26,10)	16	(2)(3)(11)	NO	PROPOSITION 4.18, 4.19, 4.20, or THEOREM 4.27 or 4.30
6.	(703,27,1)	26	(19)(37)	NO	THEOREM 4.4
7.	(352,27,2)	25	(32)(11)	NO	Ch. 5, Problem 19
8.			(2)(16)(11)	NO	THEOREM 5.21
9.			(4)(8)(11)	NO	THEOREM 5.21
0.			$(2)^2(8)(11)$	NO	THEOREM 5.21
1.			$(2)(4)^2(11)$	NO	THEOREM 5.21
2.			$(2)^3(4)(11)$	NO	THEOREM 5.21
3.			$(2)^5(11)$	NO	THEOREM 5.21
4.	(79,27,9)	18	(79)	NO	THEOREM 4.5
5.	(55,27,13)	14	(5)(11)	NO	THEOREM 4.4
6.	(757,28,1)	27	(757)	YES	PG(2,27)
7.	(253,28,3)	25	(11)(23)	NO	Ch.5, Problem 13
8.	(127,28,6)	22	(127)	NO	THEOREM 5.3
9.	(109,28,7)	21	(109)	YES	Ch.4, Example 7
10.	(85,29,9)	19	(5)(17)	NO	THEOREM 4.4
11.	(64,28,12)	16	(64)	NO	THEOREM 4.30 or 4.31

	(v,k,λ)	n	GROUP	EXIST ?	EXAMPLE OR NONEXISTENCE PROOF
112.			(2)(32)	NO	THEOREM 4.33
113.			(4)(16)		
114.			$(2)^2(16)$	YES	Ch.4, Example 4½
115.			(8)(8)		
116.			(2)(4)(8)	YES	Ch.4, Example 4½
117.			$(2)^3(8)$	YES	Ch.4, Example 4½
118.			$(4)^3$	YES	Ch.4, Example 4
119.			$(2)^2(4)^2$	YES	Ch.4, Example 4
120.			$(2)^4(4)$	YES	Ch.4, Example 4
121.			$(2)^6$	YES	Ch.4, Example 4
122.	(813,29,1)	28	(3)(271)	NO	THEOREM 5.3
123.	(407,29,2)	27	(11)(37)	NO	THEOREM 4.4
124.	(204,29,4)	25	(4)(3)(17)	NO	PROPOSITION 4.19
125.			$(2)^2(3)(17)$	NO	PROPOSITION 4.18 OR 4.19
126.	(59,29,14)	15	(59)	YES	H(59)
127.	(871,30,1)	29	(13)(67)	YES	PG(2,29)
* 128.	(291,30,3)	27	(3)(97)	NO	THEOREM 4.4

	(v,k,λ)	n	GROUP	EXIST ?	EXAMPLE OR NONEXISTENCE PROOF
9.	(175,30,5)	25	(25)(7)	NO	PROPOSITION 4.30
0.			$(5)^2(7)$	YES	Ch.4, Example 4
1.	(311,31,3)	28	(311)	NO	THEOREM 5.4
2.	(156,31,6)	25	(4)(3)(13)	YES	PG(3,5)
3.			$(2)^3(3)(13)$	NO	PROPOSITION 4.18 OR 4.19
4.	(63,31,15)	16	(9)(7)	YES	PG(5,2)
5.			$(3)^2(7)$	YES	Ch.4, Example 8
6.	(993,32,1)	31	(3)(331)	YES	PG(2,31)
7.	(249,32,4)	28	(3)(83)	NO	THEOREM 4.5
8.	(1057,33,1)	32	(7)(151)	YES	PG(2,32)
9.	(529,33,2)	31	(529)	NO	THEOREM 5.3
0.			$(23)^2$	NO	THEOREM 4.41
1.	(265,33,4)	29	(5)(53)	NO	THEOREM 4.4
2.	(177,33,6)	27	(3)(59)	NO	THEOREM 5.6
3.	(133,33,8)	25	(7)(19)	YES	Ch.4, Problem 10
4.	(97,33,11)	22	(97)	NO	THEOREM 4.4
5.	(67,33,16)	17	(67)	YES	H(67)

	(v,k,λ)	n	GROUP	EXIST ?	EXAMPLE OR NONEXISTENCE PROOF
146.	(375,34,3)	31	(3)(125)	NO	Ch. 5, Problem 20
147.			(3)(5)(25)		
148.			$(3)(5)^3$		
149.	(103,34,11)	23	(103)	NO	Ch.5, Problem 5
150.	(1191,35,1)	34	(3)(397)	NO	THEOREM 4.4
151.	(239,35,5)	30	(239)	NO	THEOREM 5.4
152.	(171,35,7)	28	(9)(19)	NO	THEOREM 4.5
153.			$(3)^2(19)$	NO	THEOREM 4.5
* 154.	(120,35,10)	25	(8)(3)(5)	NO	THEOREM 4.33
* 155.			(2)(4)(3)(5)	NO	THEOREM 4.32 or 4.3
* 156.			$(2)^3(3)(5)$	NO	THEOREM 4.27 or 4.3
157.	(71,35,17)	18	(71)	YES	H(71)
158.	(1261,36,1)	35	(13)(97)	NO	THEOREM 4.4
159.	(631,36,2)	34	(631)	NO	THEOREM 5.3
160.	(421,36,3)	33	(421)	NO	THEOREM 5.3
161.	(253,36,5)	31	(11)(23)	NO	THEOREM 5.3
162.	(181,36,7)	29	(181)	NO	THEOREM 5.3
163.	(127,36,10)	26	(127)	NO	THEOREM 5.3

	(v,k,λ)	n	GROUP	EXIST?	EXAMPLE OR NONEXISTENCE PROOF
4.	(141,36,9)	27	(3)(47)	NO	THEOREM 5.6
5.	(85,36,15)	21	(5)(17)	NO	THEOREM 4.4
6.	(1333,37,10)	36	(31)(43)	NO	THEOREM 5.12
7.	(223,37,6)	31	(223)	NO	THEOREM 5.3
8.	(149,37,9)	28	(149)	NO	THEOREM 5.4
9.	(112,37,12)	25	(2)(61)	NO	THEOREM 4.5
0.	(1407,38,1)	37	(3)(7)(67)	YES	PG(2,37)
1.	(75,37,18)	19	(3)(25)	NO	THEOREM 4.4
2.			$(3)(5)^2$	NO	THEOREM 4.4
3.	(704,38,2)	36	(64)(11)	NO	THEOREM 4.30
4.			(2)(32)(11)	NO	THEOREM 4.32
5.			(4)(16)(11)	NO	THEOREM 4.32
6.			$(2)^2(16)(11)$	NO	THEOREM 4.32
7.			$(8)^2(11)$	NO	THEOREM 4.32
8.			(2)(4)(8)(11)	NO	THEOREM 4.32
9.			$(2)^3(8)(11)$	NO	THEOREM 4.32
0.			$(4)^3(11)$		
1.			$(2)^2(4)^2(11)$		

	(v,k,λ)	n	GROUP	EXIST?	EXAMPLE OR NONEXISTENCE PROOF
182.			$(2)^4(4)(11)$		
183.			$(2)^6(11)$		
* 184.	(495,39,3)	36	(9)(5)(11)	NO	PROPOSITION 4.19
185.			$(3)^2(5)(11)$	NO	PROPOSITION 4.19
186.	(79,39,19)	20	(79)	YES	H(79)
187.	(1561,40,1)	39	(7)(223)	NO	THEOREM 4.4
188.	(521,40,3)	37	(521)	NO	THEOREM 4.4
189.	(391,40,4)	36	(17)(23)	NO	THEOREM 5.4
190.	(131,40,12)	28	(131)	NO	THEOREM 4.5
191.	(121,40,13)	27	(121)	YES	PG(4,3)
192.			$(11)^2$	NO	Ch. 5, Problem 15
* 193.	(105,40,15)	25	(3)(5)(7)	NO	THEOREM 4.27 or 4.
194.	(1641,41,1)	40	(3)(547)	NO	THEOREM 4.4
195.	(411,41,4)	37	(3)(137)	NO	THEOREM 4.4
196.	(329,41,5)	36	(7)(47)	NO	THEOREM 5.4
197.	(165,41,10)	31	(3)(5)(11)	NO	Ch. 5 Problem 21
198.	(83,41,20)	21	(83)	YES	H(83)
199.	(1723,42,1)	41	(1723)	YES	PG(2,41)

	(v,k,λ)	n	GROUP	EXIST ?	EXAMPLE OR NONEXISTENCE PROOF
0.	(575,42,3)	39	(25)(23)	NO	THEOREM 4.4
1.			$(5)^2(23)$	NO	THEOREM 4.4
2.	(288,42,6)	36	(32)(9)	NO	THEOREM 4.30
3.			(2)(16)(9)	NO	THEOREM 4.32
4.			(4)(8)(9)	NO	THEOREM 4.32
5.			$(2)^2(8)(9)$	NO	THEOREM 4.32
6.			$(2)(4)^2(9)$	NO	THEOREM 4.32
7.			$(2)^3(4)(9)$	NO	THEOREM 4.32
8.			$(2)^5(9)$	NO	THEOREM 4.32
9.			$(32)(3)^2$	NO	THEOREM 4.30
0.			$(2)(16)(3)^2$	NO	THEOREM 4.32
1.			$(4)(8)(3)^2$		
2.			$(2)^2(8)(3)^2$		
3.			$(2)(4)^2(3)^2$	NO	THEOREM 4.32
4.			$(2)^3(4)(3)^2$	NO	THEOREM 4.32
5.			$(2)^5(3)^2$	NO	THEOREM 4.32
6.	(259,43,7)	36	(7)(37)	NO	THEOREM 4.5

	(v,k,λ)	n	GROUP	EXIST ?	EXAMPLE OR NONEXISTENCE PROOF
217.	(87,43,21)	22	(3)(29)	NO	THEOREM 4.4
218.	(1893,44,1)	43	(3)(631)	YES	PG(2,43)
219.	(1981,45,1)	44	(7)(283)	NO	THEOREM 5.12
220.	(991,45,2)	43	(991)	NO	THEOREM 5.3
221.	(221,45,9)	36	(13)(7)	NO	THEOREM 5.20
222.	(199,45,10)	35	(199)	NO	THEOREM 4.4
* 223.	(111,45,18)	27	(3)(37)	NO	THEOREM 4.30
224.	(100,45,20)	25	(4)(25)	NO	THEOREM 4.33
225.			$(2)^2(25)$	NO	THEOREM 4.30
226.			$(4)(5)^2$		
227.			$(2)^2(5)^2$	NO	Ch.4, Problem 21
228.	(91,45,22)	23	(7)(13)	NO	THEOREM 4.4
229.	(2071,46,1)	45	(19)(109)	NO	THEOREM 5.12
230.	(691,46,3)	43	(691)	NO	THEOREM 5.3
231.	(415,46,5)	41	(5)(83)	NO	Ch. 5, Problem 16
232.	(231,46,9)	37	(3)(7)(11)	NO	Ch. 5, Problem 22
* 233.	(208,46,10)	36	(16)(13)	NO	THEOREM 4.30

	(v,k,λ)	n	GROUP	EXIST ?	EXAMPLE OR NONEXISTENCE PROOF
4.			(2)(8)(13)	NO	THEOREM 4.32
5.			$(4)^2(13)$		
6.			$(2)^2(4)(13)$		
7.			$(2)^4(13)$		
8.	(139,46,15)	31	(139)	NO	THEOREM 5.3
9.	(95,47,23)	24	(5)(19)	NO	THEOREM 4.4
0.	(2257,48,1)	47	(37)(61)	YES	PG(2,47)
1	(1129,48,2)	46	(1129)	NO	THEOREM 4.4
2.	(565,48,4)	44	(5)(113)	NO	THEOREM 4.4
3.	(189,48,12)	36	(27)(7)	NO	THEOREM 4.30
4.			(3)(9)(7)		
5.			$(3)^3(7)$		
6.	(2353,49,1)	48	(13)(181)	NO	THEOREM 5.12
7.	(1171,49,2)	47	(1171)	NO	THEOREM 5.3
8.	(785,49,3)	46	(5)(157)	NO	THEOREM 4.4
9.	(589,49,4)	45	(19)(31)	NO	THEOREM 4.5
0.	(393,49,6)	43	(3)(131)	NO	THEOREM 5.3

	(v,k,λ)	n	GROUP	EXIST ?	EXAMPLE OR NONEXISTENCE PROO
251.	(337,49,7)	42	(337)	NO	THEOREM 4.4
252.	(295,49,8)	41	(5)(59)	NO	THEOREM 5.3
253.	(197,49,12)	37	(197)	YES	Ch.4, Example 7
254.	(169,49,14)	35	(169)	NO	THEOREM 4.4
255.			$(13)^2$	NO	THEOREM 4.4
256.	(113,49,21)	28	(113)	NO	THEOREM 4.4
257.	(99,49,24)	25	(9)(11)	NO	THEOREM 4.39
258.		25	$(3)^2(11)$	YES	Ch.4, Example 8
259.	(2451,50,1)	49	(3)(19)(43)	YES	PG(2,49)
260.	(491,50,5)	45	(491)	NO	Ch.5, Problem 17
261.	(351,50,7)	43	(27)(13)	NO	THEOREM 4.4
262.			(3)(9)(13)	NO	THEOREM 4.4
263.			$(3)^3(13)$	NO	THEOREM 4.4
* 264.	(176,50,14)	36	(16)(11)	NO	THEOREM 4.30
265.			(2)(8)(11)	NO	THEOREM 4.32
266.			$(4)^2(11)$		
267.			$(2)^2(4)(11)$		

	(v,k,λ)	n	GROUP	EXIST?	EXAMPLE OR NONEXISTENCE PROOF
8.			$(2)^4(11)$		

In order to verify Conjecture 6.2 for $k \leq 75$, we must ıeck that there does not exist an abelian (v,k,λ)-difference ɜt in G in which the Sylow p-subgroup of G is cyclic and p ivides n. There are 56 cases to consider of which 34 are ɪrked by asterisks in Table 6-1. The remaining 22 cases ɔpear in Table 6-2 with references to a nonexistence proof.

ABLE 6-2

(v,k,λ)	n	GROUP	EXIST?	NONEXISTENCE PROOF
(171,51,15)	36	(9)(19)	NO	THEOREM 4.30
(160,54,18)	36	(32)(5)	NO	THEOREM 4.30
(441,56,7)	49	(49)(9)	NO	See [12]
		$(49)(3)^2$	NO	See [12]
(1065,57,3)	54	(3)(5)(71)	NO	THEOREM 4.4
(153,57,21)	36	(9)(17)	NO	THEOREM 4.30
(280,63,14)	49	(8)(5)(7)	NO	THEOREM 4.33
		(2)(4)(5)(7)	NO	THEOREM 4.33
		$(2)^3(5)(7)$	NO	THEOREM 4.30

	(v,k,λ)	n	GROUP	EXIST ?	EXISTENCE PROOF
10.	(2146,66,2)	64	(2)(29)(37)	NO	THEOREM 4.30
11.	(1431,66,3)	63	(27)(53)	NO	THEOREM 4.4
12.	(144,66,30)	36	(16)(9)	NO	THEOREM 4.30
13.			$(16)(3)^2$	NO	THEOREM 4.30
14.			(9)(8)(2)	NO	THEOREM 4.33
15.			$(9)(4)^2$	NO	THEOREM 4.30
16.			$(9)(4)(2)^2$	NO	THEOREM 4.30
17.			$(9)(2)^4$	NO	THEOREM 4.30
18.	(1140,68,4)	64	(4)(3)(5)(19)	NO	THEOREM 4.19
19.	(783,69,6)	63	(27)(29)	NO	THEOREM 4.30
20.	(806,70,6)	64	(2)(13)(31)	NO	THEOREM 4.19
21.	(231,70,21)	49	(3)(7)(11)	NO	THEOREM 4.33
22.	(640,72,8)	64	(128)(5)	NO	THEOREM 4.30

APPENDIX A
PERMUTATION GROUPS

This appendix presents the basic terminology and a
w results concerning permutation groups which we shall
quire in the text. We only scratch the surface of a very
ch field indeed. The reader is referred to Biggs and
ite [15] and to the excellent monograph of Wielandt [141].

A permutation of a finite set X is a one-to-one
rrespondence $g: X \longrightarrow X$. Two permutations, g and h, can be
nposed to give the permutation gh, according to the rule
n)(x) = g(h(x)). Under this operation, the permutations
rm a group Sym(X), called the symmetric group on X. When
is a fixed set of cardinality n under discussion, we also
note the symmetric group on these n objects by S_n.

If G is a subgroup of Sym(X) then the pair (G,X)
said to be a permutation group of degree |X|. Or, we
y simply that G is a permutation group on X. For each
X, the set

$$Gx = \{g(x) \mid g \in G\}$$

called the orbit of x under G. The orbits partition X.
e subgroup of all elements of G which leave the element
fixed is called the stabilizer of x; it is denoted by G_x.

Lemma A.1. $|Gx| = [G:G_x]$.

Proof. Consider the following mapping from G to
: map each $g \in G$ to $g(x) \in Gx$. The map is onto Gx. More-
er elements of G have the same image if and only if they
ree on x--that is, if and only if they are in the same left
set of G_x. Hence the number of cosets of Gx in G equals
e cardinality of G_x.□

More generally, given a set $S \subseteq X$, the <u>pointwise stabilizer</u> of S is the subgroup of all elements fixing each point of S and the <u>setwise stabilizer</u> of S is the subgroup of all elements which fix S as a set--that is, permute the elements of S among themselves.

The following result relates the number of orbits o a permutation group G to the number of points fixed by each o its elements. It is usually known as Burnside's Lemma, but several authors have recently pointed out that this attributi is incorrect and that it ought to be called the Cauchy-Frobeniu Lemma.

<u>Proposition A.2</u>. <u>The number</u> t <u>of orbits of</u> G <u>on</u> X <u>is given by</u>

$$t = \frac{1}{|G|} \sum_{g \in G} |\mathrm{Fix}(g)|,$$

<u>where</u> Fix(g) <u>is the set of fixed points of</u> g.

<u>Proof</u>. Consider the number of pairs (g,x) with $g \in G$, $x \in X$ and $g(x)=x$. Counting firstly over elements of G and secondly over X, we obtain the equality

$$\sum_{g \in G} |\mathrm{Fix}(g)| = \sum_{x \in X} |G_x|.$$

Let $x_1, \ldots, x_t$ be representatives of the t orbits. If $x \in Gx_i$ and g is any element such that $g(x)=x_i$, then $G_x = g^{-1}G_{x_i}g$. In particular, $|G_x| = |G_{x_i}|$. Hence

$$\sum_{g \in G} |\mathrm{Fix}(g)| = \sum_{i=1}^{t} \sum_{x \in Gx_i} |G_{x_i}|$$

$$= \sum_{i=1}^{t} |Gx_i|\,|G_{x_i}|$$

$$= \sum_{i=1}^{t} |G|$$

$$= t|G|,$$

where the third equality is a consequence of the lemma above.□

A permutation group G is transitive on X if it has just one orbit. That is, for any $x, y \in X$, there exists an element $g \in G$ such that $g(x) = y$. Observe that, in the transitive case, the previous results assert that

$$|X| = [G : G_x] \quad \text{and} \quad |G| = \sum_{g \in G} |\mathrm{Fix}(g)|.$$

Next we consider the action of G_x on X.

Proposition A.3. Suppose that G is a transitive group on X and let r(x) be the number of orbits of G_x on X. Then

$$r(x) = \frac{1}{|G|} \sum_{g \in G} |\mathrm{Fix}(g)|^2$$

n particular, r(x) does not depend on x.

Proof. By the Cauchy-Frobenius lemma applied to G_x,

$$r(x)|G_x| = \sum_{g \in G} |\mathrm{Fix}(g)| .$$

ow the right-hand side does not depend on x. For, let y be ny other points and take h to be a permutation group carrying to x. There is a one-to-one correspondence between the points ixed by the permutation $g \in G_x$ and the points fixed by the ermutation $h^{-1}gh$ in Gy. Now, since $|G_x|$ is also independent f x, the number r(x) does not depend on x.

We have

$$r(x)|G| = |X| \; r(x)|G_x| = \sum_{x \in X} \sum_{g \in G_x} |\mathrm{Fix}(g)|.$$

nterchanging the order of summation, we have

$$r(x)|G| = \sum_{g \in G} \sum_{x \in \mathrm{Fix}(g)} |\mathrm{Fix}(g)| = \sum_{g \in G} |\mathrm{Fix}(g)|^2 . \square$$

The _rank_ of a transitive permutation group G on X is the number of orbits of G_x on X. The previous proposition shows that this is well defined.

If G is a permutation group on X then G naturally acts as a permutation group on X x X, by the rule g((x,y)) = (g(x),g(y)). Suppose that G is transitive on X. Given $\alpha \in X$, we can define a one-to-one correspondence between the orbits of G on X x X and the orbits of G_α on X. The orbit Δ of G on X x X corresponds to the set

$$\Gamma_\Delta = \Gamma_\Delta(\alpha) = \{\gamma | (\alpha,\gamma) \in \Delta\},$$

which is an orbit of G_α on X. Conversely, the orbit Γ of G_α on X corresponds to the unique orbit Δ_Γ of G on X x X which contains $\{(\alpha,\gamma) | \gamma \in \Gamma\}$. (Verify that this is a one-to-one correspondence.) An immediate consequence is the following Proposition.

Proposition A.4. _Let_ G _be a transitive permutation group on_ X. _The number of orbits of_ G _on_ X x X _equals the rank of_ G.

Groups of rank 2 are especially interesting. In this case G_x has two orbits which are necessarily $\{x\}$ and $X - \{x\}$. By Proposition A.4, there is an element of G carrying any particular ordered pair of distinct elements of X to any other.

In this spirit we make the following definition. A permutation group G on X is _k-transitive_ ($k \geq 1$) if, given any two ordered k-tuples $(x_1,\ldots,x_k),(y_1,\ldots,y_k)$ of distinct elements, there is some element g in G such that

$$g(x_i) = y_i \quad (i=1,\ldots,k).$$

Proposition A.5. _Suppose that_ G _is transitive on_ X. _Then_ G _is_ k-_transitive on_ X _if and only if_ G_x _is_ (k-1)-_transitive on_ $X-\{x\}$.

Proof. Suppose that G_x is (k-1)-transitive on {x}. Given two ordered k-tuples of distinct elements, $_1,\ldots,x_k$) and ($y_1,\ldots,y_k$), we can choose g_1, g_2 in G and in G_x such that

$$g_1(x_1) = x, \quad g_2(y_1) = x$$

$$h(g_1(x_i)) = g_2(y_i) \quad (\text{for } i=2,\ldots k)$$

inductive hypothesis. Then $g_2^{-1}hg_1$ is the desired ement carrying one k-tuple into the other.□

A group is said to be sharply k-transitive if : is k-transitive and if the subgroup stabilizing k points s the identity. Equivalently, it is sharply k-transitive f and only if there is one and only one element carrying ıy k-tuple of distinct elements into any other.

A sharply 1-transitive group is said to be regular. ecessarily $|G| = |X|$ and the identity element is the only lement fixing a point. A possibly intransitive group G is alled semiregular if it acts as a regular permutation group n each orbit--that is, if each orbit has size $|G|$.

Related to transitivity is the notion of homo- eneity. A permutation group G on X is k-homogeneous ($k \geq 1$) :, given any two sets, $\{x_1,\ldots,x_k\}$ and $\{y_1,\ldots,y_k\}$, of distinct elements, there is an element of G carrying one et to the other. A k-transitive group is clearly k-homogeneous. ıe converse is false in general, but its failure is quite ımited. Livingstone and Wagner proved that for $k>4$, -homogeneity implies k-transitivity. For $k \leq 4$, Kantor has etermined all exceptions. (See [33, p.92].)

APPENDIX B.
BILINEAR AND QUADRATIC FORMS

This appendix presents the theory of bilinear forms and quadratic forms which we shall require in Chapters 2 and 3. While the reader has undoubtedly met these topics in a course on linear algebra, the first two sections review them from, perhaps, a more abstract point of view. The abstraction facilitates proofs, in the third section, of Witt's Extension Theorem and Cancellation Theorem for quadratic forms over an arbitrary field (including the case of fields of characteristi 2.) The last section classifies nondegenerate forms over F_p (p odd).

B.1 BILINEAR FORMS

Let V be a vector space of finite dimension n over a field F. A bilinear form on V is a map $B: V \times V \longrightarrow F$ which is F-linear in each coordinate.

(1) B is said to be a symmetric bilinear form if $B(u,v) = B(v,u)$ for all $u,v \in V$. (B is also said to be an orthogonal bilinear form or a scalar product.)

(2) B is said to be an alternating bilinear form if $B(u,u) = 0$ for all $u \in V$ or, equivalently, if $B(u,v) = -B(v,u)$ for all $u, v \in V$. (B is also said to be symplectic.)

In this monograph we are only concerned with symmetric bilinear forms. (Note that an alternating form is symmetric if char F=2.) A pair (V,B), where B is a symmetric bilinear form on the vector space V, is called a scalar product space.

Every element $v \in V$ defines an F-linear functional $f_v: V \longrightarrow F$ by the rule $f_v(u)=B(u,v)$. The bilinear form B is said to be nondegenerate or nonsingular if f_v is not the zero map whenever v is not the zero vector.

A vector u is orthogonal to a vector v if $B(u,v)=0$. (Orthogonality is a symmetric relation for symmetric or alternating bilinear forms.) More generally, a subspace U is orthogonal to a subspace W if every vector in U is orthogonal to every vector in W. The perp(endicular) of a subspace U, denoted $U^{\perp}$, is the subspace consisting of all vectors orthogonal to U.

It is easy to see that B is nondegenerate if and only if $V^{\perp} = \{0\}$. In this case, we have

(1) $\dim U + \dim U^{\perp} = n$, and

(2) $(U^{\perp})^{\perp} = U$,

or all subspaces U. The map $U \longrightarrow U^{\perp}$ defined on all sub-paces, is its own inverse. This map is called the polarity efined by B.

We introduce the following notation. If U_1 and U_2 re orthogonal subspaces of V such that $U_1 \cap U_2 = \{0\}$, we rite $U_1 \perp U_2$ for the subspace which is the sum of U_1 and $_2$. The notation is simply intended to emphasize the ortho-onality.

A vector u is isotropic if $B(u,u)=0$. A subspace U s isotropic if $U \cap U^{\perp} \neq \{0\}$. It is totally isotropic (or elf-orthogonal) if $U \subseteq U^{\perp}$.

Proposition B.1. Suppose that char $F \neq 2$. A subspace is totally isotropic if and only if every vector in U is sotropic.

Proof. Suppose that every vector in U is isotropic. ıen for all $x,y \in U$, we have

$$\begin{aligned} 0 &= B(x+y,\ x+y) \\ &= B(x,x) + B(x,y) + B(y,x) + B(y,y) \\ &= 2\,B(x,y). \end{aligned}$$

nce char $F \neq 2$, then $B(x,y)=0$. Hence U is totally isotropic. e converse is clear.□

Proposition A.1 implies that any scalar product ace can be "diagonalized" in the sense of the following sult.

Theorem B.2. *Suppose that* char $F \neq 2$. *Every scalar product space* (V,B) *can be decomposed as*

$$V = \langle x_1 \rangle \perp \ldots \perp \langle x_s \rangle \perp \langle y_1 \rangle \perp \ldots \perp \langle y_r \rangle,$$

where each x_i *is isotropic and each* y_i *is not isotropic*.

Proof. Call a subspace U of V *indecomposable* if it is not the direct sum of two orthogonal subspaces. Certainly V is an orthogonal direct sum of indecomposable subspaces. The theorem will follow once we show that an indecomposable subspace must have dimension 1. Suppose that U is indecomposable. If U is totally isotropic then any direct sum decomposition of U is an orthogonal direct sum decomposition; hence dim U must equal 1. If U is not totally isotropic then, by Proposition A.1, there exists a nonisotropic vector x in U. Then $U = \langle x \rangle \perp (\langle x \rangle^{\perp} \cap U)$. Hence U must have dimension 1.□

The hypothesis that char $F \neq 2$ is necessary. Conside for example, the vector space $V = F_2^2$, with the symmetric bilinear form

$$B((x_1,x_2), (y_1,y_2)) = x_1y_2 + x_2y_1.$$

Check that every vector is isotropic, yet that B is not totally isotropic. Moreover, V cannot be diagonalized in the sense of Theorem A.2.

Let (V_1,B_1) and (V_2,B_2) be two scalar product spaces over a field F. We say that the spaces are *equivalent* if there exists an isomorphism $\sigma: V_1 \longrightarrow V_2$ such that $B_1(u,v) = B_2(\sigma(u),\sigma(v))$ for all $u,v \in V_1$. In the event that $V_1=V_2$, we say simply that the bilinear forms B_1 and B_2 are *equivalen*

Fix a basis $e_1,\ldots,e_n$ of V. The bilinear form B gives rise to a matrix, also denoted $B = (b_{ij})$, with

$$b_{ij} = B(e_i,e_j).$$

This matrix in turn determines B completely since if $x=\sum x_ie_i$ and $y=\sum y_je_j$ then

$$B(x,y) = \sum b_{ij} x_i y_j$$

we represent x and y by the row vectors $X = (x_1, \ldots, x_n)$ and $= (y_1, \ldots y_n)$ then

$$B(u,v) = XBY^T .$$

is straightforward to check that a bilinear form B is non- generate if and only if its matrix B is nonsingular. Also, bilinear forms B_1 and B_2 are equivalent if and only if associated matrices are equivalent in the sense that

$$B_1 = AB_2A^T$$

some nonsingular matrix A.

The <u>determinant</u> of the bilinear form B is the deter- ant of the associated matrix. Of course, this depends on choice of basis $e_1, \ldots, e_n$. If we change bases, det B is ltiplied by the square of the change-of-basis matrix. cordingly we define the determinant of a bilinear form only to multiplication by nonzero squares of F. Clearly if B_1 d B_2 are equivalent bilinear forms on V then det B_1 and t B_2 are the same (up to multiplication by a nonzero square.)

2 QUADRATIC FORMS

A <u>quadratic form</u> on V is a map $Q: V \longrightarrow F$ such that

(1) $Q(au) = a^2Q(u)$ for all $a \in F$ and $u \in V$, and

(2) the map $B_Q(u,v) = Q(u+v) - Q(u) - Q(v)$ is a bilinear form.

map B_Q is clearly a symmetric bilinear form; it is called e <u>scalar product associated</u> with Q. Call Q <u>nondegenerate</u> B is.

A vector u is <u>singular</u> if $Q(u)=0$. A subspace U is <u>ngular</u> if it contains a nonzero singular vector and is <u>tally singular</u> if every vector in U is singular.

If char $F \neq 2$, quadratic forms and scalar products are rtually the same notion. For, Q gives rise to B_Q, and from

B_Q we can recover Q according to the rule $Q(x)=\frac{1}{2}B(x,x)$. (So, in this case we can associate to Q the matrix $\frac{1}{2}B$.) In this case, a subspace is singular (resp., totally singular) if and only if it is isotropic (resp., totally isotropic).

If char F=2, this is no longer true. Many quadratic forms Q may give rise to the same (alternating) bilinear form B_Q. Also, totally singular and totally isotropic are not synonymous. Consider, for example, the quadratic form on $V=F_2^{\,2}$ given by $A((x_1,x_2)) = x_1x_2$. The space V is itself totally isotropic but not totally singular.

The pair (V,Q) is formally called a <u>quadratic space</u>. If (V_1,Q_1) and (V_2,Q_2) are quadratic spaces over F, we call them <u>equivalent</u> if there exists an isomorphism of vector spaces $\sigma\colon V_1 \longrightarrow V_2$ such that $Q_1(v) = Q_2(\sigma(v))$ for all $v \in V_1$. (In the event that $V_1=V_2$, we say simply that the quadratic forms Q_1 and Q_2 are <u>equivalent</u>.) We also say that V_1 and V_2 are <u>isometric</u> and we call the map σ an <u>isometry</u> from V_1 to V_2.

Let V_1 and V_2 be F-vector spaces with quadratic forms Q_1 and Q_2, respectively. Then by $V_1 \perp V_2$ we denote the vector space $V_1 \oplus V_2$ with the quadratic form $Q_1 \oplus Q_2$ such that

$$(Q_1 \oplus Q_2)(v_1 + v_2) = Q_1(v_1) + Q_2(v_2)$$

for $v_1 \in V_1$ and $v_2 \in V_2$.

Last of all, we say that a quadratic form Q on V <u>represents</u> the scalar $a \in F$ if there exists a vector $v \in V$ such that $Q(v)=a$. Equivalent quadratic forms clearly represent the same scalars.

B.3 WITT'S THEOREMS

Let Q be a quadratic form on the F-vector space V of dimension n. Let U be a subspace of V. An injection $\sigma\colon U \longrightarrow V$ is a <u>metric injection</u> if $Q(u) = Q(\sigma(u))$ for all $u \in U$. (That is, σ is an isometry from U to a subspace of V). When can σ be extended to an isometry from V to itself?

Theorem B.3. (*Witt's Extension Theorem*) *Let* Q *be a* *ndegenerate quadratic form on* V. *Let* U *be a subspace of* V. *ery metric injection* σ: U $\longrightarrow$ V *can be extended to an* *ometry of* V.

Proof. We proceed by induction on the dimension m U. The theorem is obvious for m=0. Assume that m>0 and at the theorem is true for all subspaces U of dimension ss than m. Let P be a subspace of dimension (m-1) of U. e restriction of σ to P may be extended to an isometry σ_0 V. Let $\sigma^*=\sigma_0^{-1}\sigma$. If σ^* can be extended to an isometry * of V, then σ can be extended to the isometry $\sigma_0\sigma_0^*$ of V. us, there is no loss of generality if we assume that σ aves fixed elements of P. Set $W=\sigma(U)$.

Let S be the set of subspaces with the following operty: σ may be extended to a metric injection of X+U, aving elements of X fixed. The set S is not empty, since } ε S. Let X_1 be a maximal element of S, and let σ_1 be the tric injection of $U_1=X_1+U$ into V extending σ and leaving ements of X_1 fixed. Let $W_1=\sigma_1(U_1)$ and let $P_1=X_1+P$.

If $P_1=U_1$ then σ_1 is the identity and the theorem is vious. If not, let u_1 be an element of U_1 not in P_1 and $w_1=\sigma_1(u_1)$. Then $U_1=P_1+Fu_1$ and $W_1=P_1+Fw_1$ and $Q(u_1)=Q(w_1)$.

Let B be the bilinear form associated with Q.

Lemma B.4. *Suppose that* z *and* z' *are vectors such* *at*

(1) $z \notin U_1$ *and* $z' \notin W_1$

(2) $z-z' \in P_1^{\perp}$

(3) $B(z',w_1) = B(z,u_1)$

(4) $Q(z') = Q(z)$.

en σ_1 *can be extended to a metric injection* σ_2: U_1 + Fz$\longrightarrow$ + Fz'.

Proof of Lemma. Let σ_2 be the injection of vector aces from U_1 + Fz to W_1 + Fz' such that $\sigma_2|_{U_1} = \sigma_1$ and (z) = z'. We shall see that σ_2 preserves the value of Q. x ε U_1, then $x=p+au_1$ for some p ε P_1 and a ε F. And, (x) = p + aw_1. Since B(z'-z,p) and $B(z',w_1) = B(z,u_1)$ we ve $B(z,x) = B(z',\sigma(x))$.

Now, $Q(x) = Q(\sigma_1(x))$. Hence, for all $b \in K$,

$$\begin{aligned} Q(x + bz) &= Q(x) + bB(z,x) + b^2Q(z) \\ &= Q(\sigma_1(x)) + bB(z',\sigma_1(x)) + b^2Q(z') \\ &= Q(\sigma_1(x) + bz') \\ &= Q(\sigma_2(x + bz)). \end{aligned}$$

Hence σ_2 is a metric injection.□

Let $H = \langle u_1 - w_1\rangle^{\perp}$, an (n-1)-dimensional subspace. For $z \in H$, we have $B(z,u_1) = B(z,w_1)$. Applying the lemma above with $z'=z$, we see that the maximality of X_1 implies that $z \in U_1$ or $z \in W_1$. Hence $H \subseteq U_1 \cup W_1$. And, $H = (H \cap U_1) \cup (H \cap W_1)$. But, a vector space cannot be the union of two proper subspaces. (Why?) Hence, either $H \subseteq U_1$ or $H \subseteq W_1$.

If $U_1=V$ we are done. Otherwise $\dim U_1 = \dim W_1 \leq n-1 = \dim H$. Hence either $H=U_1$ or $H=W_1$. Thus $u_1-w_1 \in H$ is orthogonal to at least one of u_1 and w_1. But since $B(u_1,u_1) = B(w_1,w_1)$ then $B(u_1, u_1-w_1) = B(w_1-u_1,w_1)$. Hence u_1-w_1 is orthogonal to both u_1 and w_1. Thus $U_1=W_1=H$.

Let z be an element of V-H. So $B(z,u_1-w_1)\neq 0$. We shall construct an element z' with the properties required in the lemma above. Then the metric injection σ_2 in the lemma will be the desired isometry of V extending σ_1.

Clearly $w_1 \notin P_1$. Therefore $P_1{}^{\perp}$ contains a vector not orthogonal to w_1 and (by multiplying by a suitable scalar) a vector x such that $B(x,w_1) = B(z,u_1-w_1) \neq 0$. Since $B(u_1-w_1,w_1) = 0$ then x is not orthogonal to u_1-w_1. That is, $x \notin H^{\perp}$. Since $x \in P_1{}^{\perp}$ and $H=U_1=P_1+Fu_1$ then $B(u,x_1)\neq 0$ and $B(z+x,u_1-w_1)=B(x,u_1)\neq 0$, which shows that z+x is not in $W_1=H$. Let c be an element of F. Since $u_1-w_1 \in W_1$, then $(z+x+c(u_1-w_1$ $\notin W_1$. Since $(u_1-w_1) \in H^{\perp}$ and $P_1 \subseteq H$, then $(z+x+c(u_1-w_1))-z \in P_1{}^{\perp}$. We have

$$\begin{aligned} Q(u_1-w_1) &= Q(u_1) + Q(w_1) - B(u_1, w_1) \\ &= 2Q(u_1) - B(u_1,w_1) \\ &= B(u_1,u_1) - B(u_1w_1) \\ &= 0. \end{aligned}$$

It follows that

$$Q(z+x+c(u_1-w_1)) = Q(z+x) + cB(z+x,u_1-w_1).$$

Since $B(z+u,u_1-w_1)\neq 0$, the scalar c may be chosen so that $Q(z+x+c(u_1-w_1))=Q(z)$. If we set $z'=z+x+c(u_1-w_1)$ then z and z' satisfy the lemma. This completes the proof.□

Witt's Extension Theorem makes a significant statement about permutation groups. The collection of all isometries from V to itself form a group known as the <u>isometry group</u> or <u>orthogonal group</u> of V. The theorem implies that this group acts transitively on nonzero singular vectors (since if x and y are nonzero singular vectors, there is certainly a metric injection $\sigma:\langle x\rangle \longrightarrow V$ mapping x to y; this map can be extended to an isometry of V).

More generally, the <u>isometry group</u> is transitive on the set of totally singular subspaces of the same dimension. Consequently, we have the following result.

<u>Corollary B.5</u>. <u>Let</u> Q <u>be a nondegenerate quadratic form on</u> V. <u>Any two maximal totally singular subspaces have the same dimension</u>.

<u>Proof</u>. Suppose that U_1 and U_2 are each maximal totally singular subspaces and suppose that $\dim U_1 < \dim U_2$. There exists a metric injection from U_1 to a subspace of U_2, which can be extended to an isometry σ of the entire space. But, then $\sigma^{-1}(U_2)$ is a totally singular subspace properly containing U_1. This contradicts the maximality of U_1. Hence $\dim U_1 = \dim U_2$.□

We can therefore define the <u>index</u> of a nondegenerate quadratic space (V,Q) to be the dimension of a maximal totally singular subspace. (Clearly the index is at most ½(dim V).)

The following result is also quite useful.

<u>Corollary B.6</u>. <u>Let</u> Q <u>be a nondegenerate quadratic form on</u> V. <u>If</u> U_1 <u>and</u> U_2 <u>are isometric subspaces then</u> $U_1^\perp$ <u>and</u> $U_2^\perp$ <u>are also isometric</u>.

<u>Proof</u>. Let $\sigma: U_1 \longrightarrow U_2$ be an isometry. Extend σ to an isometry σ* from V to itself. The restriction of σ* to $U_1^\perp$ is the desired isometry from $U_1^\perp$ to $U_2^\perp$. □

This corollary is known as Witt's Cancellation Theorem. For, suppose that (V_i, Q_i) are nondegenerate quadratic spaces (i=1,2,3,4). Corollary B.6 implies that if $V_1 \perp V_2$ is isometric to $V_3 \perp V_4$ and V_1 is isometric to V_3, then V_2 is isometric to V_4.

B.4 FINITE FIELDS OF ODD CHARACTERISTIC

In this section, we consider the structure of quadratic forms and scalar products over $\mathbb{F}_p$, where p is an odd prime. (Things are identical over $\mathbb{F}_q$, where q is an odd prime power.) Since the field has odd characteristic, quadratic forms and scalar products are essentially the same. For a change of pace, we adopt the latter point of view.

Let d be a fixed nonsquare in $\mathbb{F}_p$.

Theorem B.7. *Every nondegenerate scalar product B on $\mathbb{F}_p^m$ is equivalent either to*

$$x_1y_1 + \cdots + x_{m-1}y_{m-1} + x_my_m$$

or

$$x_1y_1 + \cdots + x_{m-1}y_{m-1} + dx_my_m.$$

In particular, two nondegenerate scalar products on $\mathbb{F}_p^m$ are equivalent if and only if their determinants are the same (up to multiplication by a square).

Proof. By Proposition B.1, any nondegenerate scalar product is equivalent to one of the form

$$s_1x_1y_1 + \cdots + s_mx_my_m.$$

Since the squares form a subgroup of index 2 in the multiplicative group of $\mathbb{F}_p$, every s_i is either a square or d times a square. So, we may assume that B is equivalent to

$$x_1y_1 + \cdots + x_jy_j + dx_{j+1}y_{j+1} + \cdots + dx_my_m.$$

The theorem will follow once we show that the scalar products $u_1v_1 + u_2v_2$ and $du_1v_1 + du_2v_2$ are equivalent. (For then we replace every two occurrences of d by two occurrences of 1.)

To see this, notice that

$$\begin{pmatrix} 1 & a \\ -a & 1 \end{pmatrix} \begin{pmatrix} 1 & 0 \\ 0 & 1 \end{pmatrix} \begin{pmatrix} 1 & a \\ -a & 1 \end{pmatrix}^T = \begin{pmatrix} 1+a^2 & 0 \\ 0 & 1+a^2 \end{pmatrix}$$

This shows that $u_1v_1+u_2v_2$ and $(1+a^2)u_1v_1 + (1+a^2)u_2v_2$ are equivalent. Now, in the field F_p, there must exist an element a such that $(1+a^2)$ is a nonsquare. (See Problem 6 of Chapter 1.)

We have now shown that every scalar product is equivalent to one of the two listed in the theorem. These scalar products are distinguished by their determinants.□

Theorem B.8. Let B be a nondegenerate scalar product on an F_p-vector space V of dimension 2n. There exists a totally isotropic subspace of dimension n if and only if $(-1)^n$ det B is a square in F_p.

Proof. Consider the scalar product

$$(x_1y_{n+1}-x_{n+1}y_1)+\ldots+(x_{n-1}y_{2n-1}-x_{2n-1}y_{n-1})+(x_ny_{2n}-dx_{2n}y_n)$$

on F_p^{2m}. There is clearly a totally isotropic subspace of dimension n-1--namely,

$$U=\{(x_1,\ldots,x_{n-1},0,\ldots,0)|x_1,\ldots,x_{n-1} \varepsilon F_p\}$$

We claim that U is a maximal totally isotropic subspace. For, consider the quotient space $W = U^\perp/U$. If U were properly contained in a totally isotropic subspace, then W would have a nonzero isotropic subpsace (namely the image of W modulo U). However,

$$U^\perp = \{(y_1,\ldots,y_n,0,\ldots,0,y_{2n})|y_1,\ldots,y_n,y_{2n} \varepsilon F_p\}$$

and the scalar product induced on $W = U^\perp/U$ is

$$\bar{x}_n\bar{y}_{2n} - d\bar{x}_{2n}\bar{y}_n,$$

where the bar denotes reduction module U. But this scalar product has no nonzero isotropic vector (since $a^2-db^2\neq 0$ for all a,b,ε F_p). Since U is therefore a maximal totally isotropic subspace, there is no totally isotropic subpsace of dimension n by Corollary B.5.

By contrast, consider the scalar product

$$(x_1y_{n+1}-x_{n+1}y_1) + \dots + (x_ny_{2n}-x_{2n}y_n).$$

The subspace

$$U'=\{(x_1,\dots,x_n,0,\dots,0) \mid x_1,\dots,x_n \in \mathbb{F}_p\}$$

is a totally isotropic subspace of dimension n. Hence one of the two inequivalent types of scalar products has a totally isotropic subspace of dimension n; the other does not. Therefore B admits such a subspace if and only if det B differs by a square from $(-1)^n$, which is the determinant of the second scalar product above--that is, if and only if $(-1)^n$ det B is a square in $\mathbb{F}_p$.□

Last of all, we mention the following result concerning the number of isotropic vectors of a scalar product.

<u>Proposition B.9</u>. (1) <u>The number of solutions in</u> $\mathbb{F}_p$ <u>of</u> $(x_1^2-x_{n+1}^2)+\dots+(x_n^2-x_{2n}^2)=0$ is $p^{2n-1}+p^n-p^{n-1}$.

(2) <u>The number of solutions in</u> $\mathbb{F}_p$ <u>of</u> $(x_1^2-x_{n+1}^2)+\dots+(x_{n-1}^2-x_{2n-1}^2) + (x_n^2-dx_{2n}^2)=0$ is $p^{2n-1}-p^n+p^{n-1}$.

(3) <u>The number of solutions in</u> $\mathbb{F}_p$ <u>of either</u> $(x_1^2-x_{n+1}^2)+\dots+(x_n^2-x_{2n}^2)-x_{2n+1}^2=0$ or $(x_1^2-x_{n+1}^2)+\dots+(x_n^2-x_{2n}^2)-dx_{2n+1}^2 = 0$ is p^{2n}.

<u>Proof</u>. (1) For convenience, set $f(n)=p^{2n-1}+p^n-p^{n-1}$. We begin with the observation that the equation

$$u^2-v^2=a$$

has 2p-1 solutions in $\mathbb{F}_p$ if a=0 and has p-1 solutions if a≠0. (Cf., Problem 6 of Chapter 1.)

This observation includes the case n=1 of the statement. We proceed by induction. Assume that the statement holds for some integer n. We must count the number of solutions to

$$(x_1^2-x_{n+2}^2)+\ldots+(x_n^2-x_{2n+1}^2) = -(x_{n+1}^2-x_{2n+2}^2). \qquad (*)$$

By inductive hypothesis, there are f(n) ways to choose x_1, ..., x_n, x_{n+2}, ..., x_{2n+1}, so that the left-hand side of (*) is zero. By the observation above, for each such choice there are (2p-1) choices for (x_{n+1},x_{2n+2}) which satisfy (*).

For each of the remaining $p^{2n}-f(n)$ choices of x_1, ..., x_n, x_{n+2}, ..., x_{2n+1}, the left-hand side of (*) is not zero. By the observation above, for each such choice there are (p-1) choices of (x_{n+1},x_{2n+2}) which satisfy (*). Thus, the number of solutions to (*) is

$$(2p-1)f(n) + (p-1)(p^{2n}-f(n)) = f(n+1).$$

This completes the inductive step.

(2) and (3) are proven similarly.□

The equality in the two cases in (3) is not an accident. For, if Q is a nonsingular quadratic form in a vector space of odd dimension, Q and dQ are inequivalent but have the same zeroes.

APPENDIX C.
INVARIANT FACTORS

This appendix presents the theory of invariant factors (also called elementary divisors) of an integral matrix, which we require in §2.3. For convenience, we also extend the notion of invariant factors to rational matrices.

Theorem C.1. Let A be an m x m matrix with integral entries. There exist integral unimodular matrices P, Q such that

(1) PAQ is a diagonal matrix, $PAQ=\mathrm{diag}(d_1,\ldots,d_m)$

(2) d_i divides d_{i+1}, for $i=1,\ldots, m-1$.

Moreover the d_i are determined up to sign and are called the invariant factors of A.

Proof. If B and C are integral m x m matrices, write $B \sim C$ if C can be obtained from B by a finite sequence of elementary row and column operations; that is, if $C=PBQ$ where P and Q are products of elementary matrices (and thus, in particular, unimodular). Also, if B is a nonzero matrix, let $\phi(B)$ be the minimum of the absolute values of the nonzero entries of B.

If A=0 there is clearly nothing to prove. So, assume that $A\neq 0$. Choose B such that $B\sim A$ and $\phi(B)$ is minimal. After possibly permuting rows and columns we may assume that b_{11} is a nonzero entry of minimal absolute value in B. Suppose first that $b_{11}\nmid b_{1j}$ for some $j>1$. By hypothesis $|b_{11}|\leq|b_{1j}|$. By the Euclidean algorithm, there exist q,r such that $b_{1j}=qb_{11}+r$ and $0<|r|<|b_{11}|$. However, if we replace the j-th column of B by itself plus q times the first column, the resulting matrix has an entry of nonzero absolute value smaller than $|b_{11}|=\phi(B)$, contradicting the choice of B. Thus

$_{11}$ divides b_{1j}. By performing the above column operations, e may assume then that $b_{1j}=0$ for $j>1$. Similarly we may ssume that $b_{j1}=0$ for $j>1$. Finally, if $b_{11}\nmid b_{ij}$ for some ,j then adding the i-th row to the first yields a matrix $\sim A$ with $c_{11}=b_{11}\nmid c_{1j}$. But then the column operations given bove will yield a contradiction to the choice of B. Hence $_{11}|b_{ij}$ for all i,j. Now ignore the first row and column of and proceed inductively on the remainder of the matrix. his proves (1) and (2). (This proof is nonconstructive in hat no algorithm is given for finding B. The reader might ike to supply such an algorithm by exploiting the sort of ow and column operations discussed above.)

Suppose next that there exist integral unimodular atrices P,Q,P',Q' such that $PAQ=D=\mathrm{diag}(d_1,\ldots,d_m)$ and $'AQ'=D'=\mathrm{diag}(d_1',\ldots,d_m')$, with the entries in each matrix atisfying the divisibility conditions in the theorem. uppose that $|d_i|=|d_i'|$ for $i=1,\ldots,j-1$ but that $|d_j|>|d_j'|$. e shall reach a contradiction.

Let M be the $\mathbb{Z}$-module spanned by the rows of D nd let G be the quotient group $\mathbb{Z}^m/M$. Clearly $G=G_1 \times \ldots \times G_m$, here G_i is a cyclic group of order d_i. Similarly, define M' o be the $\mathbb{Z}$-module spanned by the rows of D', let G' be the uotient group $\mathbb{Z}^m/M'$. Say $G=G_1' \times \ldots \times G_m'$ where G_i' is a yclic group of order d_i'. Since there exist integral nimodular matrices U and V such that $UDV=D'$, then the groups and G' are isomorphic.

Now, if H is an abelian group, let $e_d(H)$ be the umber of elements of H having order divisible by d. If H is cyclic group of order n, then $e_d(H)=|(d,n)|$. Also $e_d(H_1 \times H_2)=$ $_d(H_1)e_d(H_2)$.

We have supposed that $|d_i|=|d_i'|$ for $i=1,\ldots,j-1$ and $d_j|>|d_j'|$. Since G and G' are isomorphic we have $e_{d_j}(G)=$ $_{d_j}(G')$. Now, $e_{d_j}(G)=d_1\ldots d_{j-1}d_j\ldots d_j$. And,

$$
\begin{aligned}
e_{d_j}(G') &= |(d_j,d_1')\ldots(d_j,d_m')| \\
&= |(d_j,d_1)\ldots(d_j,d_{j-1})(d_j,d_j')\ldots(d_j,d_m')| \\
&= |d_1\ldots d_{j-1}(d_j,d_j')\ldots(d_j,d_m')|.
\end{aligned}
$$

Since $|d_j|<|(d_j,d_j')|$ and $|d_j|\leq|(d_j,d_r')|$ for $r=j+1,\ldots,m$, we have

$$e_{d_j}(G') < |d_1\ldots d_{j-1}d_j\ldots d_j| = e_{d_j}(G).$$

The contradiction show that the invariant factors are determined up to sign.□

We can expand the notion of invariant factors to an arbitrary rational matrix A. If x and y are rational numbers say that x divides y if y=dx for some integer d. With this definition, we have the following result.

Theorem C.2. *Let* A *be an* m x m *matrix with rational entries. There exist integral unimodular matrices* P,Q *such that*

(1) PAQ *is a diagonal matrix*, $PAQ=\mathrm{diag}(d_1,\ldots,d_m)$

(2) d_i *divides* d_{i+1} for $i=1,\ldots,m$.

Moreover, the d_i *are determined up to sign and are called the invariant factors of* A.

Proof. Let N be an integer such that NA has integral entries. Apply Theorem C.1 to obtain the invariant factors $d_1,\ldots,d_m$ of NA. The invariant factors of A are simply $d_1/N,\ldots,d_m/N$.□

APPENDIX D.
REPRESENTATION THEORY

This appendix introduces the representation theory of finite groups. After defining representations in general, we merely scratch the surface by treating in depth only the case of abelian groups. The methods for studying symmetric designs developed in the text can be used profitably with non-abelian groups (see, for example, Lander [81]) but it would take us too far afield to develop the ordinary and modular representation theory of nonabelian groups. The reader is referred to [30].

The first five sections of this appendix should be read before §3.3; the final section may be postponed until before §4.6.

D.1 INTRODUCTION

Let G be a finite group and let V be a vector space over a field K. A representation of G with representation module V is a homomorphism ρ: $G \longrightarrow Aut(V)$, where Aut(V) is the group of all invertible linear transformations from V to itself. The degree of the representation is the dimension n of V.

Similarly, a matrix representation of G of degree n is a homomorphism $\bar{\rho}$: $G \longrightarrow GL(n,K)$, where GL(n,K) is the group of all n x n invertible matrices with entries in K. It is clear that once we fix a basis for V, we have a one-to-one correspondence between representations ρ with representation module V and matrix representations $\bar{\rho}$ of degree n.

Examples. (1) Let G be a permutation group acting on an n-element set Ω. Let V be the K-vector space consisting of all formal sums $\sum_{\omega \in \Omega} a_\omega \omega$ with addition and scalar multiplication performed componentwise; Ω is a basis for V. We can define a representation ρ by specifying how each $\rho(g)$ acts on the basis Ω: Let $\rho(g)$ map each element ω to $g\omega$. For obvious reasons, ρ is called a permutation representation. (The associated matrix representation, with respect to the basis Ω, maps each element $g \in G$ to the appropriate permutation matrix.)

(2) The most important permutation representation of a group G occurs when Ω consists of the elements of G, with the group G acting by left multiplication. This representation of degree $|G|$ is called the left regular representation of G.

(3) The trivial representation of G on any vector space V is defined by letting $\rho(g)$ be the identity map of V, for all $g \in G$. With respect to any basis, the associated permutation representation $\overline{\rho}$ takes each element to the identity matrix.

(4) Let G be a cyclic group of order n, with generator g. Let K be a field containing a primitive n-th root of unity, ζ. Let V be a one-dimensional vector space over K. We have a representation ρ_1 with representation module V given by

$$\rho_1(g^i)\colon v \longmapsto \zeta^i v,$$

for all $v \in V$ and $i=0,1,\ldots,n-1$. More generally, define ρ_j by

$$\rho_j(g^i)\colon v \longmapsto \zeta^{ij} v$$

for each $j=0,1,\ldots,n-1$.

(5) Let G be a cyclic group of prime order p, with generator g. Let K be a field of characteristic p. We have a matrix representation $\overline{\rho}$ of degree 2 given by

$$\rho(g^i) = \begin{pmatrix} 1 & i \\ 0 & 1 \end{pmatrix}$$

for $i=0,1,\ldots,p-1$.

(6) Let G be a cyclic group of order 3, generated y g. Let K be any field. There is a matrix representation of degree Z with

$$\rho(1) = \begin{pmatrix} 1 & 0 \\ 0 & 1 \end{pmatrix} \; ; \quad \bar{\rho}(g) = \begin{pmatrix} 1 & -1 \\ 1 & 0 \end{pmatrix} \; ; \quad \bar{\rho}(g^2) = \begin{pmatrix} 0 & 1 \\ -1 & -1 \end{pmatrix} .$$

Another useful concept is the <u>group ring</u> KG, which onsists of all formal sums $\Sigma_{g\varepsilon G}\, a_g g$, with $a_g \varepsilon K$. Addition s defined componentwise. Multiplication is defined by sing the distributive law and the group operation of G:

$$(\sum_{g\varepsilon G} a_g g)(\sum_{h\varepsilon H} b_h h) = \sum_{g\varepsilon G} \sum_{h\varepsilon H} (a_g b_h) gh.$$

learly KG is a ring with identity and is commutative if and nly if G is.

The group ring KG acts naturally on every represen- ation module V, over the field K, for G according to the rule

$$(\sum a_g g)(v) = \sum a_g(\rho(g)(v)),$$

or all $v \varepsilon V$. In this way every such representation module an be viewed as a left KG-module. Conversely, let V be a left G-module (which we henceforth assume is finite-dimensional s a K-vector space). Each element $g \varepsilon G$ gives rise to an nvertible linear transformation $\rho(g)$ defined by

$$\rho(g): v \longmapsto gv.$$

nen ρ is a representation of G. In essence, we have three quivalent points of view: representations, matrix represen- ations, and left KG-modules. We shall switch freely from ne point of view to another, according to which makes ideas ost clear. We should remark that the group ring KG is tself a left KG-module. The corresponding representation is imply the left regular representation of G over K.

Two representations ρ and σ of G over K with representation modules V and W, respectively, are equivalent if there is an isomorphism α: V $\longrightarrow$ W such that

$$\alpha\rho(g)\alpha^{-1} = \sigma(g)$$

for all $g \in G$. Of course, two matrix representations $\bar{p}$ and $\bar{\sigma}$ are called equivalent if there is a nonsingular matrix $\bar{\alpha}$ such that

$$\bar{\alpha}\bar{\rho}(g)\bar{\alpha}^{-1} = \bar{\sigma}(g)$$

for all $g \in G$. In terms of our third point of view, ρ and σ are equivalent if and only if V and W are isomorphic as KG-modules. Equivalence is an equivalence relation. We write $\rho \sim \sigma$ to indicate that ρ is equivalent to σ.

D.2 NEW REPRESENTATIONS FROM OLD

Given a representation, or two, there are a number of important ways in which to produce further representations. Let ρ_1 and ρ_2 be two representations of G over K with representation modules V_1 and V_2, respectively. (When convenient, we shall regard V_1 and V_2 as KG-modules.)

(1) the direct sum $\rho_1 \oplus \rho_2$ with representation module $V_1 \oplus V_2$ is defined in the obvious way: $(\rho_1 \oplus \rho_2)(g) = \rho_1(g)\oplus\rho_2(g) \in \mathrm{Aut}(V_1 \oplus V_2)$.

(2) Let L be an extension field of K. By extending the field of scalars, every vector space V over K gives rise to a vector space V_L over L. (Formally, this is accomplished by taking tensor products, $V_L = L \otimes_K V$.) Any element of Aut(V) gives rise to an element of Aut(V_L), by linearity. So, each representation ρ of G over K gives rise to its extension ρ_L over L.

In terms of matrix representations, this procedure is even simpler. If $\bar{\rho}$ is a matrix representation over K, then $\bar{\rho}_L$ is defined simply by regarding each matrix as being defined over the larger field.

Thinking in terms of the KG-module V, the procedure amounts simply to extending the field of scalars of V to yield a LG-module V_L.

(3) We can use automorphisms of G to generate new representations. Let ρ be a representation of G with module V and let α be an automorphism of G. We define a representation ρ^{α}, called the <u>twist</u> of ρ by α, by the rule

$$\rho^{\alpha}(g) = \rho(\alpha(g)).$$

Check that this is indeed a representation.

(4) We can also make use of field automorphisms of K. To do this we start with a matrix representation $\bar{\rho}$ of G. If σ is an automorphism of the field K, we define a matrix representation $\bar{\rho}^{\sigma}$, called the <u>twist</u> of ρ by σ, by the rule

$$\bar{\rho}^{\sigma}(g) = \sigma(\bar{\rho}(g)),$$

where σ is applied to each element of the matrix $\bar{\rho}(g)$. Check that this is indeed a matrix representation.

(5) Last of all, we define the <u>contragredient</u> ρ^* of a representation ρ. Suppose that ρ is a representation of G with module V. Consider the dual vector space $V^* = \mathrm{Hom}_K(V,K)$ consisting of all linear functionals on V. We define ρ^* to be a representation of G with module V^* as follows. For each $g \in G$, we define $\rho^*(g)\colon V^* \longrightarrow V^*$ such that $(\rho^*(g))(f)$ is the linear functional such that

$$[(\rho^*(g))(f)](v) = f[\rho(g^{-1})(v)]$$

for all $v \in V$. Check that ρ^* is indeed a representation.

Notice that in this manner V^* acquires the structure of a left KG-module. We call $V^* = \mathrm{Hom}_K(V,K)$, with this structure, the <u>contragredient</u> module of V.

Contragredients are quite simple to describe in terms of matrix representations. For

$$\bar{\rho}^*(g) = (\bar{\rho}(g)^{-1})^T,$$

the transpose inverse. (To see why this is so, think in terms of a dual basis of V.)

In the text we shall be interested in the twists or contragredients of various representations, or modules. We should note that the twist of a sum of representations is isomorphic to the sum of the twists and similarly for contragredients.

Sometimes it will happen that a representation module is isomorphic to its contragredient. In this case, we say that the module is self-contragredient. Self-contragredient modules will play an important role in §3.3 and later sections.

D.3 IRREDUCIBLE AND INDECOMPOSABLE REPRESENTATIONS

A representation $\rho: G \longrightarrow \mathrm{Aut}(V)$ is reducible if, as a KG-module, V has some nontrivial sub-KG-module (that is, a sub-KG-module other than 0 or V). Otherwise, we say that V is simple and ρ is irreducible. Recall the examples above.

(1) and (2) A permutation representation of degree greater than 1 is always reducible, since the element $\sum_{\omega \in \Omega} \omega$ spans a one-dimensional subspace closed under the action of G.

(3) A trivial representation of degree greater than one is reducible, since every subspace is closed under the action of G.

(4) Since a one-dimensional KG-module is necessarily simple, the fourth example is irreducible.

(5) Suppose that the matrices of the fifth example act on the vector space of column vectors. Then $(1,0)^T$ spans a one-dimensional subspace fixed by G.

(6) Whether the sixth example is reducible depends on the choice of the field K. We can find a one-dimensional submodule closed under the action of G if and only if K contains an eigenvalue of $\rho(g)$ and $\rho(g^2)$. (Why?) The eigenvalues are $\frac{1}{2}(1 \pm \sqrt{-3})$. Thus p is reducible over fields containing a square root of -3, such as the complex numbers, F_3 or F_7, and is irreducible over fields without such an element such as the rational numbers or F_5.

Motivated by this last example, we say that a representation ρ over K is *absolutely irreducible* if ρ_L is irreducible for any extension L over K.

Irreducible representations correspond to simple KG-modules and vice versa. Let V be a KG-module (which we continue to assume is finite-dimensional as a K-vector space). There exists a *composition series* for V--that is, a chain

$$\{0\} = V_0 \subseteq V_1 \subseteq \cdots \subseteq V_{t-1} \subseteq V_t = V$$

of KG-modules such that (V_i/V_{i-1}) is a nonzero simple KG-module. (Just keep squeezing in modules whenever a quotient s not a simple module.) We say that the simple modules $(V_1/V_0),\ldots,(V_t/V_{t-1})$ are the *composition factors* of V. In his sense, the irreducible representations are the building locks for all representations. The Jordan-Holder Theorem sserts that the compositions factors depend only on the module , not on the choice of composition series. (We omit the roof since the reader has probably seen something quite imilar for groups.)

A concept related to irreducibility is indecomposibility. A representation ρ with representation module V is *ecomposable* if we can write $V = W_1 \oplus W_2$ for two nontrivial G-modules of V. Otherwise, ρ is *indecomposable*. Every ecomposable representation is clearly reducible. The converse in general false. A module V may be reducible (that is, ssess submodules M) but not be decomposable (that is, not ve any modules N such that $V = M \oplus N$). Consider the fifth ample above: if it were decomposable (as a sum of one-mensional submodules necessarily) then the matrices given uld be diagonalized. However the matrices cannot be agonalized over a field of characteristic p. (Prove this.)

Irreducibility and indecomposability do coincide, wever, provided that the characteristic of the field does t divide the order of the group.

Theorem D.1. (Maschke's Theorem). Let G be a finite oup and suppose that K is a field whose characteristic does t divide the order of G. Every indecomposable KG-module irreducible.

Proof. Let V be a KG-module. Suppose that V is reducible, containing a nontrivial sub-KG-module M. We shall show that V is decomposable. Let π be the canonical projection $\pi\colon V \longrightarrow V/M$. Suppose that we could find a map $\eta\colon V/M \longrightarrow V$, which is a KG-homomorphism, such that $\pi \circ \eta = id_{V/M}$. Then the image of η would be a sub-KG-module of V such that $V = M \oplus \text{Image}(\eta)$. (Check this.) We shall find such an η.

Now thinking of V and V/M as simply K-vector spaces it is easy to find a map $\tau\colon V/M \longrightarrow V$ such that $\pi \circ \tau = id_{V/m}$. However, there is no reason to suppose that τ is a KG-homomorphism--that is, that $\tau(gx) = g\tau(x)$ for all $g \in G$ and $x \in V/M$. The trick is to take the "average" of the maps $x \longmapsto g^{-1}\tau(gx)$. Define η by the rule

$$\eta(x) = \frac{1}{|G|} \sum_{g \in G} g^{-1}\,\tau(gx)$$

Then η is a K-linear map from V/M to V such that

$$\eta(hx) = \frac{1}{|G|} \sum_{g \in G} g^{-1}\tau(ghx) = \frac{1}{|G|} \sum_{k \in G} hk^{-1}\tau(kx) = h\eta(x),$$

where $k=gh$. Thus η is the appropriate KG-homomorphism. The only requirement of the proof is that $|G|$ have an inverse in K.□

(Maschke's Theorem is sharp in the sense that if char K divides $|G|$ then there must exist an indecomposable, but reducible, representation. We shall not prove this, however.)

The following result is an immediate consequence of the proof above of Maschke's Theorem.

Corollary D.2. Let V be a KG-module and suppose that the hypotheses of Maschke's Theorem hold. Every composition factor of V is actually a direct summand of V. Thus, V is isomorphic to a direct sum of its composition factors.

Since irreducible representations, or simple modules, e the fundamental building blocks of representation theory, would be useful to know whether there are finitely many of em and, if so, how to enumerate them all. The next propo- tion supplies the answer.

Proposition D.3. Every simple KG-module is iso- rphic to a composition factor of KG. In particular, there e only finitely many simple KG-modules, up to isomorphism.

Proof. Let M be a simple KG-module and let x be a nzero element of M. Define a map $f\colon KG \longrightarrow M$ by $f(a) = ax$ r all $a \in KG$. Since the image of f is a nonzero sub-KG- dule of the simple module M, the image must equal M. Thus is onto and

$$M \simeq (KG/\mathrm{Ker}\ f).$$

at is, M is a composition factor of KG. By the Jordan- lder Theorem, KG has only finitely many (at most $|G|$) mposition factors.□

When the hypotheses of Maschke's Theorem apply, the xt result follows at once from the corollary above.

Proposition D.4. Let G be a group and let K be a eld of characteristic not dividing the order of G. Then KG n be decomposed as a direct sum of simple left sub-KG-modules

$$KG = N_1 \oplus \ldots \oplus N_r$$

ery simple KG-module is isomorphic to one of the N_i.

Since KG is a sum of simple modules in this case, is said that the group ring K is semisimple. We should te that a left sub-KG-module of KG is just a left ideal; us, the N_i are the minimal left ideals of KG.

When the characteristic of K divides the order of Maschke's Theorem is false and KG is not semisimple. In is case the representation theory of G is considerably re complicated.

In the next two sections we shall discuss the repre- ntation theory of an abelian group G in the semisimple case, arting with the situation of a "large" field (one containing ough roots of unity to make everything easy) and then moving the situation of a "small" field.

D.4 ABELIAN GROUPS: SEMISIMPLE CASE

Let G be an abelian group of exponent μ and let K be a field containing a primitive μ-th root of unity, ζ. (Notice that this necessarily requires that the characteristic of K does not divide μ.) A <u>character</u> of G is a homomorphism χ from G to the multiplicative group of K. For example, the <u>trivial character</u> (or <u>principal character</u>), denoted χ_0, is the map such that $\chi_0(g) = 1$ for all $g \in G$. It is not difficult to determine all characters of G. For, decompose G as a product of cyclic groups, $G = G_1 x \ldots x G_s$, where G_i is generated by the element g_i. A character χ must carry each g_i to a $|G_i|$-th root of unity, and conversely χ is completely determined by knowing to which root of unity each of the g_i is carried. Hence, there are precisely $|G| = |G_1| x |G_2| x \ldots x |G_s|$ characters of G. (In fact the characters form a group under the rule $(\chi\eta)(g) = \chi(g)\eta(g)$ which is isomorphic to G itself. This is easy to see if G is cyclic and can be proven by induction on s, otherwise.)

We should remark that characters can easily be extended to be maps from the group ring KG onto K and it is sometimes convenient to do so. If χ is a character of G, simply let

$$\chi\left(\sum_{g \in G} a_g g\right) = \sum_{g \in G} a_g \chi(g).$$

Every character χ defines a representation ρ_χ of G with a one-dimensional representation module which we denote V_χ. We have

$$\rho_\chi(g)\colon v \longmapsto \chi(g)v.$$

for all $v \in V_\chi$. Since ρ has degree one, it is of course an absolutely irreducible representation. The associated matrix representation (with respect to any basis) is

$$\bar{\rho}_\chi(g) = (\chi(g)).$$

So, we have found $|G|$ distinct irreducible representation of G, all having degree one. By Proposition D.3

and the Jordan-Holder Theorem, these must be *all* of the composition factors of KG. Or, in other words, every simple KG-module must be isomorphic to one of the V_χ.

By Proposition D.4, the group ring can be written as a direct sum of simple (left) ideals isomorphic to the V_χ. That is,

$$KG = \bigoplus_{\chi} N_\chi$$

where the sum ranges over all characters and where N_χ is a (left) ideal isomorphic as a KG-module to V_χ. In particular, we should note that each composition factor of KG occurs with multiplicity one.

It turns out that we can identify the ideals N_χ explicitly, by giving a generator of N_χ in KG.

Theorem D.5. Let G *be an abelian group of exponent* μ *and let* K *be a field containing* μ*-th roots of unity. Then*

$$KG = \bigoplus_{\chi} KGe_\chi$$

where the sum is taken over all characters and where the elements e_χ *are given by*

$$e_\chi = \frac{1}{|G|} \sum_{g \in G} \chi(g^{-1})g.$$

Proof. We must check that the left KG-module KGe_χ is isomorphic to V_χ. This is true because for all $h \in G$

$$\begin{aligned} he_\chi &= \frac{1}{|G|} \sum_{g \in G} (g^{-1})hg \\ &= \frac{1}{|G|} \sum_{g \in G} \chi(g^{-1}h^{-1})\chi(h)hg \\ &= \chi(h)\Big[\frac{1}{|G|} \sum_{k \in G} \chi(k^{-1})k\Big] \\ &= \chi(h)e_\chi . \end{aligned}$$

This completes the proof.□

The elements e_χ turn out to have particularly nice properties. It turns out that $(e_\chi)^2 = e_\chi$ (whence the e_χ are called <u>idempotents</u>) and $e_\chi e_\eta = 0$ if $\chi \neq \eta$ (whence they are said to be <u>mutually orthogonal</u>). Both of these facts follow at once from the following proposition.

<u>Proposition D.6</u>. <u>Let</u> $G^\times$ <u>denote the group of characters of</u> G (<u>with values in</u> K).

(i) For all $h \in G$,

$$\sum_{\chi \in G^\times} \chi(h) = \begin{cases} |G| & \text{if } h \text{ is the identity element,} \\ 0 & \text{otherwise.} \end{cases}$$

(ii) For all $\chi \in G^\times$,

$$\sum_{h \in G} \chi(h) = \begin{cases} |G| & \text{if } \chi = \chi_0 \\ 0 & \text{otherwise.} \end{cases}$$

<u>Proof</u>. (i) Let $S(h)$ denote the sum in question. If h is the identity element then $S(h)$ certainly equals $|G|$. So, let h be a nonidentity element. Choose a character η such that $\eta(h) \neq 1$. Then

$$S(h) = \sum_{\chi \in G^\times} \chi(h) = \sum_{\chi \in G^\times} (\chi\eta)(h) = \eta(h) \sum_{\chi \in G^\times} \chi(h) = \eta(h) S(h).$$

Hence $S(h) = 0$. In a similar fashion, (ii) follows. □

The matrix $C = (c_{\chi g})$, whose rows are indexed by the characters and whose columns are indexed by the elements of G, with

$$c_{\chi g} = \chi(g)$$

is called the <u>character table</u> of G. The previous proposition states simply that

$$CC^T = C^T C = |G| I.$$

Proposition D.7. Let v *and* w *be two elements of the group ring* KG. *If* $\chi(v)=\chi(w)$ *for all characters* χ, *then* $v=w$.

Proof. It is enough to show that if $\chi(v)=0$ for all characters χ then $v=0$. So, suppose that $v = \sum a_g g$ and that $\chi(v)=0$ for all characters χ. Now, for all $g \in G$,

$$a_g = \frac{1}{|G|} \sum_{\chi \in G^\times} \chi(g^{-1}v),$$

by Proposition D.6 (i). Since $\chi(g^{-1}v)=\chi(g^{-1})\chi(v)=o$, then $a_g=0$. Hence $v=0$.□

In the text we shall frequently consider ideals, or ubmodules, of the group ring KG. We have already identified he simple ideals. It is not much harder to classify all the deals of KG. Let M be an ideal of KG. We say that M *nvolves* the character χ if $Me_\chi \neq 0$; since Me_χ is a submodule f the simple module KGe_χ then we must actually have $e_\chi=KGe_\chi=N_\chi$.

Proposition D.8. Suppose that S *is the set of haracters which an ideal* M *of* KG *involves. Then*

$$M = \bigoplus_{\chi \in S} N_\chi$$

n particular, an ideal of KG *is completely determined by the et of characters which it involves.*

Proof. Observe that $1=\sum e_\chi$, the sum taken over all haracters. Then

$$M=(\sum e_\chi)M = \bigoplus_{\chi \in G^\times} Me_\chi = \bigoplus_{\chi \in S} Me_\chi = \bigoplus_{\chi \in S} N_\chi,$$

ich proves the proposition.□

Last of all, we consider the subject of twists and ntragredients of representations of G over K. Let ρ_χ be e representation associated with the character χ. The ader should verify the following.

(1) Let α be an automorphism of G. The twist $(\rho_\chi)^\alpha$ is equivalent to $\rho_{(\chi^\alpha)}$, where χ^α is the twist of χ by α,

$$\chi^\alpha(g) = \chi(\alpha(g)).$$

(2) Let σ be an automorphism of K. The twist $(\rho_\chi)^\sigma$ is equivalent to $\rho_{(\chi^\sigma)}$, where χ^σ is the twist of χ by σ,

$$\chi^\sigma(g) = \sigma(\chi(g)).$$

(3) The contragredient $(\rho_\chi)^*$ of the representation ρ_χ is equivalent to ρ_{χ^*}, where χ^* is the inverse character

$$\chi^*(g) = \chi(g^{-1}) = \chi(g)^{-1}.$$

Twists and contragredients of other representations may be computed by first decomposing the representation as a sum of irreducible representations.

We should note that the twist or contragredient of a representation is an abstract representation, defined only up to equivalence. However, let N_χ be a simple ideal of KG. There is a unique ideal isomorphic to $(N_\chi)^*$, namely N_{χ^*}. We shall abuse notation slightly and speak of N_{χ^*} as _the_ contragredient of N_χ. Similarly, we speak of $N_{(\chi^\alpha)}$ as _the_ twist of N_χ by α(where α is an automorphism of G or of K). Of course, there is no need to restrict ourselves to simple ideals: if M is any ideal of KG, there is a unique ideal equivalent to the twist or the contragredient of M.

The following proposition is immediate.

Proposition D.9. Let M be an ideal of KG.

(i) _If_ α _is an automorphism of_ G, _then_ M _is its own twist by_ α _if and only if the set of characters which_ M _involves is closed under twisting by_ α.

(ii) _If_ σ _is an automorphism of_ K, _then_ M _is its own twist by_ σ _if and only if the set of characters which_ M _involves is closed under twisting by_ σ.

(iii) M *is its own contragredient if and only if the set of characters which* M *involves is closed under inverses*.

D.5 ABELIAN GROUPS: SEMISIMPLE CASE (Continued)

Let G be an abelian group of exponent μ. Suppose that F is a field which does not contain a primitive μ-th root of unity. Provided that the characteristic of F does not divide the order of G (which we shall henceforth assume is the case), there exists a finite-dimensional Galois extension K of F containing a primitive μ-th root of unity.

We already understand the representation theory of G over K, from the previous section. Now, consider the Galois group $\Gamma = \mathrm{Gal}\,(K,F)$. The characters of G (defined with values in K) fall naturally into algebraic conjugacy classes, $C_1,\dots,C_r$, where C_i contains all characters of the form $(\chi_i)^\sigma$ for some fixed character χ_i and any $\sigma \in \Gamma$. Consider the element $f_i \in KG$ defined by

$$f_i = \sum_{\chi \in C_i} e_\chi = \sum_{\chi \in C_i} \Big(\frac{1}{|G|} \sum_{g \in G} \chi(g^{-1})g\Big)$$

$$= \frac{1}{|G|} \sum_{g \in G} \Big(\sum_{\chi \in C_i} \chi(g^{-1})\Big)g.$$

All of the coefficients are left fixed by Γ, whence they all lie in F. Thus f_i lies in FG. Since the e_χ are mutually orthogonal idempotents, so are the f_i. That is, $(f_i)^2=f_i$ and $f_i f_j=0$ if $i\neq j$. We have the following result.

Theorem D.10. *Let* G *be an abelian group of exponent* μ, *let* F *be a field of characteristic not dividing the order of* G *and let* $f_1,\dots,f_r$ *be defined as above. Then*

$$FG = \bigoplus_{i=1}^{r} FGf_i$$

and the ideals FGf_i *are simple* FG-*modules*.

Proof. We need only to verify that $U=FGf_i$ is a simple FG-module. Suppose not and let W be a proper sub-FG-module of U. Consider the KG-modules $W_K \subseteq U_K$ obtained by extension of scalars from W and U, respectively. Since W is an FG-module, W_K is closed under twisting by elements of Γ. By Proposition D.9, the set of characters which W_K involves must form a union of algebraic conjugacy classes. Now, the set of characters which U_K involves is a conjugacy class. Hence either $W_K=U_K$ or $W_K=(0)$. In the latter case, $W=(0)$. In the former case, $(0)\simeq(U_K/W_K)\simeq(U/W)_K$, whence $U/W\simeq(0)$ and thus $U\simeq W$. Hence U is a simple module.□

Remark. Since the f_i are idempotents, each of the modules FGf_i is closed under multiplication; hence FGf_i is a ring with unit element f_i. Any ideal of the ring FGf_i would necessarily be a sub-FG-module of FGf_i. Since FGf_i is a simple FG-module, it has no nontrivial ideals; that is, FGf_i is a field. Thus Theorem D.10 asserts that FG can be decomposed as a direct sum of fields.

Just as before, the modules FGf_i form a complete list of all simple FG-modules (up to isomorphism). In particular, we should note that FG has the property that every composition factor occurs with multiplicity one.

Let M be an ideal of FG. We say that M involves the character χ (defined over K) if M_K involves χ . We clearly have the following result.

Proposition D.11. Let M be an ideal of FG. The set of characters which M involves is a union of algebraic conjugacy classes. The ideal M is completely determined by the set of characters which it involves.

If M is an ideal of FG, there is (as for KG), a unique ideal isomorphic to the contragredient of M or to a twist of M by any particular automorphism of F or G. We again slightly abuse notation by calling this particular ideal the contragredient or the twist, as the case may be.

We shall be concerned in the text with the situation in which $F=\mathbb{F}_p$, the field with p elements. We have the following useful result.

Proposition D.12. Let G *be an abelian group of order relatively prime to* p. *Let* α *be the automorphism of* G *given by* $\alpha(g)=g^p$, *for* $g \in G$.

If M *is any ideal of* F_pG *then* M *is its own twist under* α.

Proof. Let K be an extension field of F_p containing $|G|$-th roots of unity. Consider the ideal M_K of KG and let $\sigma \in \mathrm{Gal}\ (K,F_p)$ be the automorphism of K such that $\sigma(x)=x^p$ for $x \in K$. By the facts following Proposition D.8, the twist of M_K under α is the same as the twist of M_K under σ. But M_K is its own twist under σ, since $\sigma \in \mathrm{Gal}\ (K,F_p)$.□

Proposition D.13. Let G *be an abelian group of order relatively prime to* p. *Suppose that for some integer* f,

$$p^f \equiv -1 \text{ (mod exponent of G)}.$$

Then every simple F_pG*-module is isomorphic to its contragredient*.

Proof. Let M be any ideal in F_pG and let α be the automorphism of G given by $\alpha(g)=g^p$ for $g \in G$. Now, M is its own twist under α, and hence also its own twist under α^f. But α^f sends each element of G to its inverse. Hence M is its own contragredient (by the facts following Proposition D.8).□

.6 ABELIAN GROUPS: NON-SEMISIMPLE CASE

As a rule, it is much harder to classify all submodules of $F_p[G]$ when p divides the order of the abelian group . An important exception, however, is the case in which the ylow p-subgroup of G is cyclic.

Proposition D.14. Let P *be a cyclic group of prime* rder p^a. *For* $0 \leq d \leq p^a$, *the group ring* $F_p[P]$ *has a unique ideal* d *of* F_p*-dimension* d.

Proof. Let g be a generator of P. Observe the somorphism

$$F_p[P] \simeq F_p[X]/(X^{p^a}-1) = F_p[X]/(X-1)^{p^a}.$$

Every ideal of $R = F_p[X]/(X-1)^{p^a}$ is the image of an ideal of $F_p[X]$ under the canonical projection. If $f(x) \in F_p[X]$ is a polynomial relatively prime to $(x-1)$, then the image of $f(x)$ is a unit in the quotient ring. (Why?) More generally, if $f(x)=(x-1)^s g(x)$, with $g(x)$ relatively prime to $(x-1)$, then $(f(x))$ and $((x-1)^s)$ have the same image in R--which is an ideal of dimension p^a-s. Hence, there is a unique ideal V_d in $F_p[P]$ of dimension d--namely, $V_d=(g-1)^{p^a-d}$.

Remarks. (1) The group ring $F_p[P]$ has a unique composition series as an $F_p[P]$-module (and is therefore said to be uniserial); each composition factor is a one-dimensional trivial representation.

(2) The proof above works for any field K of characteristic p. Consequently, we note that the only ideals of $K[P] = K \otimes F_p[P]$ are $K \otimes V_d$. (By $\otimes$, we understand tensor product of F_p-modules.)

(3) Let P' be the unique subgroup of P of order p and let $\mu : [P] \longrightarrow P/P'$ be the canonical projection. Then μ induces a homomorphism $\mu : F_p[P] \longrightarrow F_p[P/P']$. It is easy to verify that the kernel of μ is $V_{p^{a-1}(p-1)}$.

More generally, consider an abelian group $G = P \times Q$, where P is a cyclic group of order p^a and Q is a group of order relatively prime to p. The group ring $F_p[G]$ is isomorphic as an $F_p[G]$-module to $F_p[P] \otimes F_p[Q]$ (simply map $(p,q) \longmapsto p \otimes q$ and extend by linearity). We can determine the ideals of $F_p[G]$ from our knowledge of the ideals of $F_p[P]$ and $F_p[Q]$.

By using the idempotents $f_1, \ldots, f_s$ determined in the last section, we can split $F_p[Q]$ as a sum of rings,

$$F_p[Q] = \bigoplus_{i=1}^{s} U_i$$

where in fact $U_i = F_p[Q]f_i$ has the structure of a field (of characteristic p). Hence, we can decompose $R = F_p[P] \otimes F_p[Q]$ as

$$F_p[P] \otimes F_p[Q] = (F_p[P] \otimes U_1) \oplus \ldots \oplus (F_p[P] \otimes U_s).$$

Every ideal of R must then be a direct sum of ideals of the $\mathbb{F}_p[P] \otimes U_i$. Since U_i is a field, any ideal of $\mathbb{F}_p[P] \otimes U_i$ must have the form $V_{d_i} \otimes U_i$, by the second remark above. Hence, all ideals of R have the form

$$\bigoplus_{i=1}^{s} (V_{d_i} \otimes U_i).$$

Such an ideal is generated by the elements $((g-1)^{p^a-d_1}, f_1), \ldots, ((g-1)^{p^a-d_s}, f_s)$ $P \times Q$.

APPENDIX E.
CYCLOTOMIC FIELDS

The field $Q(\zeta_d)$, where ζ_d is a primitive d-th root of unity, is called a _cyclotomic field_. We shall require two results about cyclotomic fields, both of which are typically proven in a first course on algebraic number theory. We state these without proof, since it would take us too far afield to review this subject here and since fortunately there are many excellent references to recommend. See, e.g., [17,118,123,139].

Let ϕ be the Euler ϕ-functions; $\phi(d)$ is the number of positive integers less than or equal to d and relatively prime to d.

Theorem E.1. The field $Q(\zeta_d)$ _is a normal extension of_ Q _having degree_ $\phi(d)$. _For each integer_ t _relatively prime to_ d, _there is a field automorphism_ α_t _of_ $Q(\zeta_d)$ _over_ Q _which carries_ ζ_d _to_ ζ_d^t. _There are_ $\phi(d)$ _distinct such automorphisms, which comprise the Galois group_ $Gal(Q(\zeta_d),Q)$.

Theorem E.2. Let p _be a prime integer_.

(1) _If_ $d=p^e$, _the ideal_ (p) in $Q(\zeta_d)$ _decomposes as_ $(p) = \pi^{\phi(d)}$, _where_ π _is the principal prime ideal_ $(1-\zeta_d)$

(2) _If_ $(d,p) = 1$, _the ideal_ (p) in $Q(\zeta_d)$ _decomposes as_ $(p) = \pi_1 \ldots \pi_g$, _where the_ π_i _are distinct prime ideals. Furthermore_, $g = \phi(d)/f$, _where_ f _is the order of_ p _modulo_ d. _The field automorphism_ α_p, _which sends_ ζ_d to ζ_d^p, _fixes every ideal_ $\pi_i (i=1,\ldots,g)$.

(3) _Suppose that_ $d=p^e d'$, _with_ $(p,d') = 1$. _Then the ideal_ (p) _in_ $Q(\zeta_d)$ _decomposes as_

$$(p) = (\pi_1 \ldots \pi_g)^{\phi(p^e)}$$

where the π_i are distinct prime ideals. Furthermore g=$\phi(d')/f$, where f is the order of p modulo d'. Let t be an integer such that (t,p) = 1 and $t \equiv p \pmod{d'}$. The field automorphism α_t, which sends ζ_d to ζ_d^t, fixes every ideal π_i (i=1,...,g).

Corollary E.3. Let $d=p^e d'$, with (p,d') = 1. Suppose that $p^j \equiv -1 \pmod{d'}$, for some integer j. Then every prime ideal dividing (p) in $Q(\zeta_d)$ is fixed under complex conjugation.

Proof. By part (3) of the previous theorem, every prime ideal dividing (p) in $Q(\zeta_d)$ is fixed by α_t provided that (t,p) = 1 and $t \equiv p^s \pmod{d'}$, for some integer s. Taking t=-1 (with s=j), we see that every ideal is left fixed by the field automorphism which carries each root of unity to its reciprocal. But, this map is simply complex conjugation.□

APPENDIX F.
P-adic NUMBERS

This appendix presents a brief introduction to the p-adic integers $\hat{\mathbb{Z}}_p$ and the p-adic field $\hat{Q}_p$. These extensions of $\mathbb{Z}$ and Q, respectively, enjoy many of the interesting properties of finite fields of characteristic p while in fact having characteristic zero. Our treatment is cursory and completely ignores any mention of the topological structure of $\hat{Q}_p$ which is quite important in its many circumstances. The interested reader is referred to [17, 123, 139].

Fix a prime p. For $n \geq 1$, let $A_n = \mathbb{Z}/p^n\mathbb{Z}$, the ring of integers (mod p^n). There is a natural projection ϕ_n from A_n to A_{n-1}, with kernel $p^{n-1}A_n$; these projections give rise to a sequence

$$\cdots \xrightarrow{\phi_{n+1}} A_n \xrightarrow{\phi_n} A_{n-1} \xrightarrow{\phi_{n-1}} \cdots \xrightarrow{\phi_3} A_2 \xrightarrow{\phi_2} A_1.$$

The <u>ring of</u> p-<u>adic integers</u>, denoted $\hat{\mathbb{Z}}_p$, consists of all sequences $x = (\ldots, x_n, \ldots, x_1)$ with

$$x_n \in A_n \quad \text{for } n \geq 1$$

$$\text{and} \quad \phi_n(x_n) = x_{n-1} \text{ for } n \geq 2.$$

(Every element x_n of A_n can be written uniquely as $x_n = \sum_{i=0}^{n-1} a$ with $a_i \in \{0,1,\ldots,p-1\}$. So, a p-adic integer can be viewed as a sum $x = \sum_{i=0}^{\infty} a_i p^i$ with $a_i \in \{0,1,\ldots,p-1\}$.)

With addition and multiplication defined component-wise, $\hat{\mathbb{Z}}_p$ forms a commutative ring with unit element $1 = (\ldots,1,\ldots 1)$. The integers $\mathbb{Z}$ are naturally embedded in $\hat{\mathbb{Z}}_p$.

Let $\varepsilon_n : \hat{\mathbb{Z}}_p \longrightarrow A_n$ be the map sending each p-adic integer x to its n-th component x_n.

Proposition F.1. *A p-adic integer x is a multiple* ɔf p^n *if and only if* $\varepsilon_n(x)=0$. *In other words, the kernel of* ε_n *is* $p^n\hat{\mathbb{Z}}_p$. *Hence* $\hat{\mathbb{Z}}_p/p^n\hat{\mathbb{Z}}_p \simeq A_n = \mathbb{Z}/p^n\mathbb{Z}$.

Proof. Clearly $p^n\mathbb{Z} \subseteq \operatorname{Ker} \varepsilon_n$. Suppose that $x = (\ldots,x_n,\ldots,x_1) \in \operatorname{Ker} \varepsilon_n$. Since $x_m \equiv 0 \pmod{p^n}$ for all $m > n$, there exists a unique element y_{m-n} of A_{m-n} such that $x_m = p^n y_{m-n}$. Then $y = (\ldots,y_m,\ldots,y_1) \in \hat{\mathbb{Z}}_p$ and $x = p^n y$. Hence $\operatorname{Ker} \varepsilon_n \subseteq p^n\mathbb{Z}$.□

Proposition F.2. (i) *A p-adic integer has a multiplicative inverse if and only if it is not a multiple of* p.

(ii) *Every nonzero p-adic integer can be written uniquely in the form* $p^n u$, *with* u *an invertible element and* $n \geq 0$.

(iii) $\hat{\mathbb{Z}}_p$ *is an integral domain*.

(iv) $\hat{\mathbb{Z}}_p$ *has a unique maximal ideal*, $p\hat{\mathbb{Z}}_p$. *All nonzero ideals of* $\hat{\mathbb{Z}}_p$ *have the form* $p^n\hat{\mathbb{Z}}_p$.

Proof. (i) Suppose that $x = (\ldots,x_n,\ldots,x_1)$ is not a multiple of p. Since x_n is not a multiple of p in A_n, there is a unique $y_n \in A_n$ such that $x_n y_n = 1$. Then $y = (\ldots,y_n,\ldots,y_1)$ is the desired inverse for x. (Since $\phi(y_n)$ is an inverse for x_{n-1}, we must have $\phi(y_n)=y_{n-1}$.) An invertible element is called a *p-adic unit*. A multiple of p cannot be a unit (since, for example, it is in the kernel of a nonzero homomorphism, ε_1).

(ii) Let $x = (\ldots,x_n,\ldots,x_1) \neq 0$. There is a largest integer n such that $x_n=0$. The p^n divides x and $x = p^n u$, with u a p-adic unit. The decomposition is clearly unique.

(iii) In view of (ii), it is enough to check that p is not a zero divisor. Suppose that $px=0$ for some

$x = (\ldots, x_n, \ldots, x_1)$. Then $px_{n+1} \equiv 0$ for all n, whence x_{n+1} has the form $p^n y_{n+1}$ for some $y_{n+1} \in A_{n+1}$. Since $x_n = \phi(p^n y_{n+1})$, we see that $x_n = 0$.

(iv) This is immediate from (ii).□

The field of fractions of $\hat{\mathbb{Z}}_p$ is called the field of p-adic numbers, denoted $\hat{Q}_p$. It suffices to provide an inverse for p, in order to obtain a field; that is $\hat{Q}_p = \mathbb{Z}_p[p^{-1}]$. Every element of $\hat{Q}_p$ can be written uniquely as $p^n u$, with u a p-adic unit and $n \in \mathbb{Z}$.

We shall be concerned in Chapter 5 with the structure of the group rings $\hat{Q}_p[G]$ and $\hat{\mathbb{Z}}_p[G]$, where G is an abelian group of order n, not divisible by p. Since every ring $\mathbb{Z}/p^n\mathbb{Z}$ is a quotient ring of $\hat{\mathbb{Z}}_p$, this will also provide us with information about the group ring $\mathbb{Z}/p^n\mathbb{Z}[G]$. It turns out that the structure of these group rings exactly mirrors that of $F_p[G]$, which fact is a direct consequence of the close analogy between cyclotomic extension of $\hat{Q}_p$ and cyclotomic extensions of F_p.

Theorem F.3. Suppose that n is an integer relatively prime to p and that $f = \mathrm{ord}_p(n)$.

(1) Let ζ be a primitive n-th root of unity in some algebraic extension of F_p. Then $K = F_p(\zeta)$ is a Galois extension of F_p of degree f. The Galois group is cyclic, generated by the automorphism α such that $\alpha(\zeta) = \zeta^p$.

(2) Let ζ be a primitive n-th root of unity in some algebraic extension of $\hat{Q}_p$. Then $\hat{K} = \hat{Q}_p[\zeta]$ is a Galois extension of $\hat{Q}_p$ of degree f. The Galois group is cyclic, generated by the automorphism α such that $\alpha(\zeta) = \zeta^p$.

We defer the proof. The work of §D.4 and §D.5 depended only on the structure of the appropriate cyclotomic extensions and so all the definitions and results concerning group rings carry over at once from $F_p[G]$ to $\hat{Q}_p[G]$. We define the $e_\chi \in \hat{Q}_p[\zeta][G]$ and $f_i \in \hat{Q}_p[G]$ as in Appendix D.

We have

$$\hat{Q}_p[G] = \bigoplus_{i=1}^{r} \hat{Q}_p[G]f_i$$

where the modules $V_i = \hat{Q}_p[G]f_i$ are simple $\hat{Q}_p[G]$-modules. Every submodule M of $\hat{Q}_p[G]$ has the form

$$M = \bigoplus_{i \varepsilon S} V_i$$

where $S \subseteq \{1,\ldots,r\}$. And, since the modules U_i are closed under twisting by the automorphism $\alpha: g \longmapsto g^p$ for $g \varepsilon G$, every submodule of $\hat{Q}_p[G]$ has this property also.

The situation is almost the same for $\hat{\mathbb{Z}}_p[G]$. The elements e_χ in fact lie in $\hat{\mathbb{Z}}_p[\zeta][G]$ (since $|G|$ has an inverse in $\hat{\mathbb{Z}}_p$) and the f_i lie in $\mathbb{Z}_p[G]$. Hence we can use the idempotents f_i to split $\hat{\mathbb{Z}}_p[G]$:

$$\hat{\mathbb{Z}}_p[G] = \bigoplus_{i=1}^{r} \hat{\mathbb{Z}}_p[G]f_i.$$

Unlike the situation for $\hat{Q}_p$, the modules $U_i = \hat{\mathbb{Z}}_p[G]f_i$ are not simple modules; however, neither are they very complicated.

Lemma F.4. *The only nonzero submodules of* $U_i = \hat{\mathbb{Z}}_p[G]f_i$ *are of the form* $p^\alpha U_i$ *for some integer* $\alpha \geq 0$.

Consequently, every submodule M *of* $\hat{\mathbb{Z}}_p[G]$ *has the form*

$$M = \bigoplus_{i \varepsilon S} p^{\alpha_i} U_i$$

where $S \subseteq \{1,\ldots,s\}$ *and* $\alpha_i \geq 0$.

We defer the proof. An immediate consequence of the lemma is the conclusion that every submodule of $\hat{\mathbb{Z}}_p[G]$ is left fixed by the automorphism $\alpha: g \longmapsto g^p$ for $g \varepsilon G$. We shall use this result in Chapter 5.

It remains only to prove Theorem F.3 and Lemma F.4. We begin with an important lemma.

Lemma F.5 (Hensel's Lemma). Let $f(x)$ be a polynomial in $\hat{\mathbb{Z}}_p[x]$ and let $\overline{f}(x) \in \mathbb{F}_p[x]$ be the reduction of $f(x) \pmod p$. Suppose that $\overline{f}$ can be factored in $\mathbb{F}_p[x]$ as

$$\overline{f} = \overline{g}\overline{h}$$

where $(\overline{g},\overline{h}) = 1$ and $\overline{g}$ is monic. Then f can be factored as $f=gh$, where g and h reduce to $\overline{g}$ and $\overline{h} \pmod p$, respectively, and where g is monic.

Proof. The lemma says that we may "lift" factorizations of $\overline{f}(x)$ (into relatively prime factors) to factorizations of $f(x)$ in $\hat{\mathbb{Z}}_p[x]$. We shall prove it by constructing this lifting one step at a time. Let $m = \deg f$ and $r = \deg \overline{g}$ (so that $\deg \overline{h} \leq m-r$). We will inductively define polynomials $g_n, h_n \in (\mathbb{Z}/p^n\mathbb{Z})[x]$ (for $n \geq 1$) such that

(1) the reduction of g_n and $h_n \pmod{p^{n-1}}$ is g_{n-1} and h_{n-1}, respectively.
(2) g_n is monic.
(3) $\deg h_n \leq m-r$.
(4) $f = g_n h_n \pmod{p^n}$.

Begin by letting $g_1=\overline{g}$ and $h_1=\overline{h}$. Suppose as inductive hypothesis that we have polynomials g_{n-1} and h_{n-1} satisfying (1) - (4) above. Let g_n' and h_n' be any polynomials in $(\mathbb{Z}/p^n\mathbb{Z})[x]$ which reduce to and have the same degrees as g_{n-1} and h_{n-1}, respectively, and such that g_n' is monic. We shall set $g_n = g_n' + p^{n-1}y_n$ and $h_n = h_n' + p^{n-1}z_n$, where y_n and z_n are polynomials to be determined in order to force $f \equiv g_n h_n \pmod{p^n}$. Now,

$$\begin{aligned} f - g_n h_n &\equiv f - g_n' h_n' - p^{n-1}(g_n' z_n + h_n' y_n) + p^{2n-2} y_n z_n \pmod{p^n} \\ &\equiv p^{n-1} a_n - p^{n-1}(g_n' z_n + h_n' y_n) \pmod{p^n} \\ &\equiv p^{n-1}(a_n - (g_n' z_n + h_n' y_n)) \pmod{p^n} \end{aligned}$$

where we can choose an appropriate polynomial a_n, since $f - g_n'h_n' \equiv f - g_{n-1}h_{n-1} \equiv 0 \pmod{p^{n-1}}$. The condition that $f \equiv g_n h_n \pmod{p^n}$ will be satisfied if

$$a_n \equiv g_n' z_n + h_n' y_n \qquad \pmod{p}.$$

Since the reductions (mod p) of g_n' and h_n' are relatively prime in $\mathbb{F}_p[x]$, we can find suitable polynomials y_n and z_n. In fact, y_n may be taken to have degree less than r. (For, if deg $y_n \geq r$, we can write $y_n = g_{n-1}q + y_n^*$, where deg y_n^* Replace z_n by $z_n + q$ and y_n by y_n^*.) Now, by inductive hypothesis $f - g_n'h_n' = p^{n-1}a_n$ has degree at most m. Also deg $h_n'y_n \leq m$. Hence also deg $g_n'z_n \leq m$. Thus deg $z_n \leq m-r$. The induction is complete.

The factors of f(x) in $\hat{\mathbb{Z}}_p[x]$ are obtained as follows. If $g_n(x) = \sum_{j=0}^{r} c_{nj} X^j$, then set $g(x) = \sum_{j=0}^{r} c_j X^j$ where for $1 \leq j \leq r$

$$c_j = (\ldots, c_{nj}, \ldots, c_{1j}) \in \hat{\mathbb{Z}}_p.$$

)efine h(x) similarly.□

Proof of Theorem F.3. (1) consists of well-known .nformation about finite fields. Since the multiplicative ;roup of a finite field is cyclic, the smallest extension ield containing a primitive n-th root of unity is $\mathbb{F}_{p^f}$, hich has degree f over $\mathbb{F}_p$. Every cyclotomic extension is a alois extension (since if $\zeta \in K$ then so are all algebraic onjugates of ζ, which are necessarily other n-th roots of nity). The automorphism α has order f and hence generates he full Galois group.

(2) is proven by considering the factorization of (x) = $x^n - 1 \in \hat{\mathbb{Z}}_p[x]$. Suppose that $x^n - 1 = d_1(x)d_2(x)\ldots d_s(x)$, here the $d_i(x) \in \hat{\mathbb{Z}}_p[x]$ are monic and irreducible. The egree of $\hat{K}$ over $\hat{Q}_p$ is the largest of the degrees of the d_i. ow, by Hensel's Lemma and the fact that $x^n - 1$ has no ultiple roots over $\mathbb{F}_p$,

$$\overline{f}(x) = x^n-1 = \overline{d}_1(x)\overline{d}_2(x)\ldots\overline{d}_s(x)$$

must be a factorization of $\overline{f}(x)$ into irreducible polynomials in $\mathbb{F}_p[x]$. The degree of K over $\mathbb{F}_p$ is again the largest degree of the d_i. Hence $[\hat{K}: \hat{Q}_p] = [K: \mathbb{F}_p] = f$.

Lastly, we determine the Galois group of $\hat{K}$ over $\hat{Q}_p$. Let ζ be a primitive n-th root of unity; an automorphism is completely determined by specifying its action on ζ. We shall show that there is an automorphism α carrying ζ to ζ^p; such an automorphism necessarily has order f and hence generates the full Galois group. Suppose that $d_1(x)$ is the minimal polynomial of ζ. Then ζ^p is a root of $d_1(x^k)$, where k is a positive integer such that $kp \equiv 1 \pmod{n}$. We must have

$$d_1(x^k) \equiv d_j(x) \qquad \pmod{x^n-1}$$

for some $1 \leq j \leq s$. To show that ζ^p is an algebraic conjugate of ζ it is necessary and sufficient to show that $j=1$. Reducing (mod p), we have

$$\overline{d}_1(x^k) \equiv \overline{d}_j(x) \qquad \pmod{x^n-1}.$$

Since $\mathrm{Gal}(K,\mathbb{F}_p)$ has an automorphism carrying roots of unity to their p-th powers, we have

$$\overline{d}_j(x) \equiv \overline{d}_1(x^k) \equiv \overline{d}_1(x) \qquad \pmod{x^n-1}.$$

In view of the correspondence between factorization of x^n-1 over $\hat{Q}_p$ and over $\mathbb{F}_p$ and the fact that x^n-1 has no multiple factors, we have $j=1$.□

<u>Proof of Lemma F.4</u>. Consider a nonzero ideal M of $U_i = \hat{\mathbb{Z}}_p[G]f_i$. After dividing by a sufficiently high power of p, we may assume that M contains an element x not divisible by p. The lemma will be established once we show that there exists an element $y \in \hat{\mathbb{Z}}_p[G]$ such that $xy=f_i$. The method is completely analogous to that used to prove Hensel's Lemma. We find, by induction, elements $y_1, y_2, \ldots$ such that

$y_j \equiv y_{j+1} \pmod{p^j}$ and $xy_j \equiv f_j \pmod{p^j}$. For $j=1$, the existence of such a y_1 follows from the simplicity of $\mathbb{F}_p[G]\bar{f}_i$ (where the bar denotes reduction mod p). The inductive step is straightforward and is left to the reader.□

REFERENCES

[1] Ahrens, R.W. & Szekeres, G. (1969). On a combinatorial generalization of 27 lines associated with a cubic surface. J.Austral.Math.Soc., 10, 485-492.

[2] Alltop, W.O. (1972). An infinite class of 5-designs. J.Com Theory, (A)12, 390-395.

[3] Anstee, R., Hall, Jr., M. & Thompson, J.G. (1980). Planes of order 10 do not have a collineation of order 5 J.Comb.Theory, (A)29, 39-58.

[4] Assmus, Jr., E.F. & van Lint, J.H. (1979). Ovals in projective designs. J. Comb.Theory, (A)27, 307-324

[5] Assmus, Jr., E.F. & Maher, D.P. (1978). Nonexistence proo for projective designs. Amer.Math.Monthly, 85, 110-112.

[6] Assmus, Jr., E.F. & Mattson, H.F. (1969). New 5-designs. J.Comb.Theory, 6, 122-151.

[7] Assmus, Jr., E.F., Mezzaroba, J.A. & Salwach, C.J. (1977). Planes and biplanes. In Higher Combinatorics, ed. M.Aigner, pp.205-212. Dordrecht: D.Reidel.

[8] Assmus, Jr., E.F. & Sachar, H.E. (1977). Ovals from the points of view of coding theory. In Higher Combinatorics, ed. M.Aigner, pp.213-216. Dordrecht: D. Reidel.

[9] Assmus, Jr., E.F. & Salwach, C.J. (1979). The (16,6,2) designs. Internat.J.Math.and Math.Sci., 2, 261-

[10] Aschbacher, M. (1981). On collineations of symmetric bloc designs. J.Comb.Theory, 11, 272-281.

[11] Baer, R. (1947). Projectivities of finite projective pl Amer.J.Math., 69, 653-684.

[12] Baumert, L.D. (1969). Difference sets. SIAM J.Appl.Mat 17, 826-833.

[13] Baumert, L.D. (1971). Cyclic difference sets. In Lecture Notes in Mathematics, 182, New York: Springer-Verlag.

[14] Beker, H. (1976). An orbit theorem for designs. Geom. Dedicata, 5, 425-433.

[15] Biggs, N.L. & White, A.T. (1979). Permutation groups and combinatorial structures, London Math.Soc.Lecture Notes Series, 33, Cambridge University Press.

[16] Block, R.E. (1965). Transitive groups of collineations of certain designs. Pac.J.Math., 15, 13-19.

[17] Borevich, Z.I. & Shafarevich, I.R. (1966). Number Theory. New York: Academic Press.

[18] Bose, R.C., Shrikhande, S.S. & Singhi, N.M. (1977). Edge-regular multigraphs and partial geometric designs. Teorie Combinatorie (Tomo I), Accad.Naz.Lincei, Rome.

[19] Bruck, R.H. (1955). Difference sets in a finite group. Trans.Amer.Math.Soc., 78, 464-481.

[20] Bruck, R.H. & Ryser, H.J. (1949). The nonexistence of certain finite projective planes. Canad.J.Math., 1, 88-93.

[21] Buekenhout, F. (1969). Une caractérisation des espaces affines basée sur la notion de droit. Math.Z., 111, 367-371.

[22] Buekenhout, F. & Shult, E. (1974). On the foundations of polar geometry. Geom.Dedicata, 3, 155-170.

[23] Burnside, W. (1955). Theory of Groups of Finite Order. (2nd ed. reprint). New York: Dover.

[24] Cameron, P.J. (1973). Extending symmetric designs, J.Comb. Theory, (A)14, 215-220.

[25] Cameron, P.J. & van Lint, J.H. (1980). Graphs, Codes and Designs. London Mathematical Society Lecture Notes in Mathematics 43. Cambridge University Press.

[26] Cameron, P.J. (1980). External results and configuration theorems for Steiner systems. Disç.Math. 7, 43-63.

[27] Cameron, P.J. (1982). Dual polar spaces, Geom.Dedicata 12, 75-86.

[28] Cameron, P.J. & Seidel, J.J. (1973). Quadratic forms over GF(2). Indag.Math., 35, 1-8.

[29] Chowla, S. & Ryser, H.J. (1950). Combinatorial problems. Canad.J.Math., 2, 93-99.

[30] Curtis, C.W. & Reiner, I. (1962). Representations of Finit Groups and Associative Algebras. New York: Interscience.

[31] Delsarte, P., Goethals, J.-M. & Seidel, J.J. (1971). Ortho gonal matrices with zero diagonal II. Canad.J.Ma 23, 816-832.

[32] Dembowski, P. (1958). Verallgemeinerungen von Transitivätskl endlicher projectiver Ebenen, Math.Z., 69, 59-89.

[33] Dembowski, P. (1968). Finite Geometries. Berlin-Heidelber New York: Springer-Verlag.

[34] Dembowski, P. & Wagner, A. (1960). Some characterizations of finite projective spaces. Arch.Math., 11, 465

[35] Denniston, R.H.F. (1976). Some new 5-designs. Bull.London Math.Soc., 8, 263-267.

[36] Dillon, J.F. (1974). Elementary Hadamard difference sets. Ph.D. Thesis, University of Maryland.

[37] Evans, T.A. & Mann, H.B. (1951). On simple difference set Sankhya 11, 357-364.

[38] Feit, W. (1970). Automorphisms of symmetric balanced incom block designs. Math.Z., 118, 40-49.

[39] Feit, W. & Thompson, J.G. (1963). Solvability of groups o odd order. Pac.J.Math., 13, 755-1029.

[40] Geramita, A.U. & Seberry, J. (1979). Orthogonal designs, quadratic forms and Hadamard matrices. In Lectu Notes in Pure and Applied Mathematics, 45 Marcel Dekker. New York: Basel.

[41] Goethals, J.-M. & Delsarte, P. (1968). On a class of majo logic decodable cyclic codes. IEEE Trans.Inform Theory, 14, 182-188.

[42] Goethals, J.-M. & Seidel, J.J. (1967). Orthogonal matric with zero diagonal, Canad.J.Math., 19, 1001-101

[43] Gordon, B., Mills, W.H. & Welch, L.R. (1962). Some new d ference sets, Canad.J.Math., 14, 614-625.

[44] Graham, R.L. & MacWilliams, F.J. (1966). On the number o information symbols in difference-set cyclic co Bell System Tech.J., 45, 1057-1070.

[45] Hadamard, J. (1893). Resolution d'une question relative aux determinants. Bull. des Sciences Math., 77, 240-246.

[46] Haemers, W. (1979). Eigenvalue techniques in design and graph theory. Thesis, Technical University of Eindhoven.

[47] Haemers, W. & Shrikhande, M. (1979). Some remarks on subdesigns of symmetric designs. J.Stat.Planning and Inference, 3, 361-366.

[48] Hall, M. Jr. (1947). Cyclic projective planes. Duke Math.J. 14, 1079-1090.

[49] Hall, M. Jr. (1967). Combinatorial Theory, Waltham, MA: Blaisdell.

[50] Hall, M. Jr. (1975). Difference sets. In Combinatorics, eds. M. Hall, Jr. & J.H. van Lint, pp. 321-346. Dordrecht: D. Reidel.

[51] Hall, M. Jr. (1980). Coding theory of designs. In London Mathematical Society Lecture Notes Series 49, eds. P.J. Cameron J.W.P. Hirschfeld & D.R. Hughes. Cambridge: Cambridge University Press.

[52] Hall, M. Jr. & Connor, W.S. (1953). An embedding theorem for balanced incomplete block designs. Canad.J. Math., 6, 35-41.

[53] Hall, M. Jr. & Ryser, H.J. (1951). Cyclic incidence matrices. Canad.J.Math., 3, 495-502.

[54] Hamada, N. (1973). On the p-rank of the incidence matrix of a balanced or partially balanced incomplete block design and its application to error-correcting codes. Hiroshima Math.J., 3, 154-226.

[55] Hamada, N. & Ohmori, H. (1975). On the BIB design having the minimum p-rank. J.Comb.Theory, (A)18, 131-140.

[56] Hardy, G.H. & Wright, E.M. (1954). An Introduction to the Theory of Numbers. Oxford: Oxford University Press, 3rd ed.

[57] Hering, C. (1979). On codes and projective designs. Kyoto Univ.Math.Res.Inst.Sem.Notes, 344.

[58] Higman, G. (1969). On the simple groups of D.G. Higman and C.C. Sims. Ill.J.Math., 13, 74-80.

[59] Hirschfeld, J.W.P. (1979). Projective Geometries Over Finite Fields. Oxford: Clarendon Press.

[60] Hughes, D.R. (1957). Collineation groups and generalized incidence matrices. Trans.Amer.Math.Soc., 86, 284-296.

[61] Hughes, D.R. (1957). Generalized incidence matrices over group algebras. Ill.J.Math., 1, 545-551.

[62] Hughes, D.R. (1957). Regular collineation groups, Proc. Amer.Math.Soc., 8, 165-168.

[63] Hughes, D.R. (1965). On t-designs and groups. Amer.J. Math., 87, 761-778.

[64] Hughes, D.R. (1978). On biplanes and semibiplanes. In Combinatorial Mathematics, eds. D.A. Holton & Jennifer Seberry. Lecture Notes in Mathematics, 686, pp. 55-58. New York: Springer-Verlag.

[65] Hughes, D.R. & Piper, F.C. (1973). Projective Planes. Berlin-Heidelberg-New York: Springer-Verlag.

[66] Ito, N. (1967). On a class of doubly, but not triply transitive permutation groups. Arch.Math. (Basel), 18, 564-570.

[67] Janko, Z. & Van Trung, T. (1981). On projective planes of order 12 with an automorphism of order 13, Part II. Geom.Ded. 11, 257-284.

[68] Janko, Z. & Van Trung, T. (1982). On projective planes of order 12 with an automorphism of order 13: Part II. Geom.Ded. 12, 87-99.

[69] Janko, Z. & Van Trung, T. (1982). The full collineation group of any projective plane of order 12 is a {2,3}-Group. Geom.Ded. 12, 101-110.

[70] Johnsen, E.C. (1964). The inverse multiplier for abelian group difference sets. Canad.J.Math., 16, 787-796

[71] Jungnickel, D. (Unpublished). Difference sets with multiplier -1.

[72] Kantor, W.M. (1969). 2-Transitive symmetric designs. Trans.Amer.Math.Soc., 146, 1-28.

[73] Kantor, W.M. (1975). Symplectic groups, symmetric designs and line ovals. J.Alg., 33, 43-58.

[74] Kantor, W.M. (1975). 2-Transitive designs. In Combinatori ed. M. Hall, Jr. & J.H. van Lint, pp. 365-418. Dordrecht: D. Reidel.

[75] Kantor, W.M. (to appear). Exponential numbers of two-weig codes, difference sets and symmetric designs.

[76] Kelly, G.S. (1979). On Automorphisms of Symmetric 2-Designs, Ph.D. Dissertation, Westfield College, University of London.

[77] Kibler, R.E. (1978). A summary of non-cyclic difference sets, k<20. J.Comb.Theory, (A)25, 62-67.

[78] Lander, E.S. (1979). On Self-Dual Codes and Projective Planes, Qualifying Dissertation, Oxford University.

[79] Lander, E.S. (1980). Topics in Algebraic Coding Theory, D.Phil. Thesis, Oxford University.

[80] Lander, E.S. (1981). Symmetric designs and self-dual codes. J.London Math.Soc., 24, 193-204.

[81] Lander, E.S. (1981). Characterization of biplanes by their automorphism groups. In Geometries and Groups, eds. M. Aigner & D. Jungnickel. Lecture Notes in Mathematics, 893, pp. 204-218. Berlin: Springer-Verlag.

[82] Lehmer, E. (1953). On residue difference sets. Canad.J. Math., 5, 425-432.

[83] Lenz, H. & Jungnickel, D. (1979). On a class of symmetric designs. Arch.Math., 33, 590-592.

[84] Luneburg, H. (1969). Transitive Erweiterungen endlicher Permutationsgruppen. Lecture Notes in Mathematics, 84. Berlin-Heidelberg-New York: Springer-Verlag.

[85] Luneburg, H. (1980). Translation planes. Berlin-New York: Springer-Verlag.

[86] MacWilliams, F.J. & Mann, H.B. (1968). On the p-rank of the design matrix of a difference set. Information and Control, 12, 474-488.

[87] MacWilliams, F.J. & Sloane, N.J.A. (1977). The theory of error-correcting codes. Amsterdam: North-Holland.

[88] MacWilliams, F.J., Sloane, N.J.A. & Thompson, J.G. (1973). On the existence of a projective plane of order 10. J.Comb.Theory, (A)14, 66-78.

[89] Mann, H.B. (1952). Some theorems on difference sets. Canad.J.Math., 4, 222-226.

[90] Mann, H.B. (1955). Introduction to algebraic number theory. Columbus, OH: Ohio State University Press.

[91] Mann, H.B. (1964). Balanced incomplete block design and abelian difference sets. Ill.J.Math., 8, 252-261.

[92] Mann, H.B. (1965). Difference sets in elementary abelian groups. Ill.J.Math., 9, 212-219.

[93] Mann, H.B. (1967). Recent advances in difference sets. Amer.Math.Monthly, 74, 229-235.

[94] Mann, H.B. (1967). Addition Theorems. New York: Interscien

[95] Mann, H.B. & McFarland, R.L. (1973). On Hadamard difference sets. In A Survey of Combinatorial Theory, eds. J.N. Srivastava et al., pp. 333-334. Amsterdam: North-Holland.

[96] McFarland, R.L. (1980). On Multipliers of Abelian Differen Sets, Ph.D. Dissertation, Ohio State University.

[97] McFarland, R.L. (1973). A family of difference sets in non-cyclic groups. J.Comb.Theory, (A)15, 1-10.

[98] McFarland, R.L. (1979). On (v,k,λ)-configurations with $v=4p^e$. Glasgow J.Math., 15, 180-183.

[99] McFarland, R.L. & Rice, B.F. (1978). Translates and multipliers of abelian difference sets, Proc.AMS, 68, 375-379.

[100] Menon, K. (1960). Difference sets in abelian groups. Pro AMS, 11, 368-376.

[101] Menon, K. (1962). On difference sets whose parameters satisfy a certain relation. Proc.AMS, 13, 739-74

[102] Neumann, P.M. (1972). Transitive permutation groups of prime degree. J.London Math.Soc., 5, 202-208.

[103] Neumann, P.M. (1978). The simplicity of the Green Heart. Seminar on Permutation Groups and Related Topics. Kyoto University.

[104] Ostrom, T.G. (1953). Concerning difference sets, Canad.J. Math., 5, 421-424.

[105] Ostrom, T.G. (1956). Double transitivity in finite projective planes. Canad.J.Math., 8, 563-567.

[106] Ostrom, T.G. & Wagner, A. On projective and affine plane with transitive collineation groups, Math.Z., 71 186-199.

[107] Ott, U. (1975). Eindliche Zyklische ebenen. Math.Z., 14 195-215.

[108] Ott, U. (1981). Some remarks on representation theory in finite geometry. In Geometries and Groups, eds. M. Aigner & D. Jungnickel. Lecture Notes in Mathematics, 893, pp. 68-110. Berlin: Springer-V

[109] Paley, R.E.A.C. (1933). On orthogonal matrices. J.Math and Physics, 12, 311-320.

[110] Parker, E.T. (1957). On collineations of symmetric designs. Proc.AMS., 8, 350-351.

[111] Petrenjuck, A.Ja. (1968). Math Zametski, 4, 417-425.

[112] Pless, V. (1972). Symmetry codes over GF(3) and new five-designs, J.Comb.Theory, 12, 119-142.

[113] Ray-Chandhuri, D.K. & Wilson, R.M. (1975). On t-designs. Osaka J.Math., 12, 737-744.

[114] Rotman, J. (1965). The Theory of Groups: An Introduction. Boston: Allyn and Bacon.

115] Ryser, H. (1963). Combinatorial mathematics. Carus Mathematical Monographs No. 14.

116] Ryser, H. (1982). The existence of symmetric block designs. J.Comb.Theory, (A)32, 103-105.

117] Salwach, C.J. (1981). Planes, biplanes and their codes. Amer.Math.Monthly, 88, 106-124.

118] Samuel, P. (1970). Algebraic Theory of Numbers. New York: Houghton-Mifflin.

119] Schmidt, W. (1976). Equations over finite fields: An elementary approach. Lecture Notes in Mathematics, 536. Berlin: Springer-Verlag.

120] Schutzenberger, M.P. (1949). A nonexistence theorem for an infinite family of symmetrical block designs. Ann.Eugenics, 14, 286-287.

121] Seberry, J. (1978). A computer listing of Hadamard matrices. Combinatorial Mathematics Proceedings, Canberra, 1977. Lecture Notes in Mathematics, 686, Berlin-Heidelberg-New York: Springer-Verlag.

122] Segre, B. (1955). Ovals in finite projective planes. Canad.J.Math., 414-416.

123] Serre, J.P. (1973). A Course in Arithmetic. New York: Springer-Verlag.

24] Serre, J.P. (1977). Linear Representations of Finite Groups. New York: Springer-Verlag.

25] Shult, E. & Yanushka, A. (1980). Near n-gons and line systems. Geom.Dedicata, 9, 1-72.

[126] Singer, J. (1938). A theorem in finite projective geometr and some applications to number theory. Trans. Amer.Math.Soc., 43, 377-385.

[127] Smith, K.J.C. (1969). On the p-rank of the incidence matrix of points and hyperplanes in a finite geometry. J.Comb.Theory, 7, 122-129.

[128] Smith, M. (1976). A combinatorial configuration associate with the Higman-Sims simple group. J.Alg., 41, 175-195.

[129] Stanton, R.G. & Sprott, D.A. (1958). A family of differe sets. Canad.J.Math., 13, 73-77.

[130] Todd, J.A. (1959). On representations of the Mathieu gro as collineation groups. J.London Math.Soc., 34, 406-416.

[131] Tsuzuku, T. (1982). Finite Groups and Finite Geometries. Cambridge Tracts in Mathematics Series No. 78. Cambridge University Press.

[132] Turyn, R. (1965). Character sums and difference sets. Pac.J.Math., 15, 319-346.

[133] Turyn, R. (1965). The nonexistence of seven difference sets. Ill.J.Math., 9, 590-594.

[134] van Lint, J.H. (1971). Coding theory. Lecture Notes in Mathematics, 201. Berlin-Heidelberg-New York: Springer-Verlag.

[135] van Lint, J.H. & Seidel, J.J. (1966). Equilateral point sets in elliptic geometry. Indag.Math., 28, 335-348.

[136] Veblen, O. & Young, J.W. (1916). Projective Geometry. Boston, MA: Ginn and Company.

[137] Wallis, W.D. (1971). Construction of strongly regular graphs using affine designs. Bull.Austral.Math Soc., 4, 41-49.

[138] Wallis, W.D., Street, A.P. & Wallis, J.S. (1972). Combi torics: Room Squares, Sum-Free Sets, Hadamard Matrices. Lecture Notes in Mathematics, 292. Berlin-Heidelberg-New York: Springer-Verlag.

[139] Weiss, E. (1963). Algebraic Number Theory. New York: McGraw-Hill.

[140] Whitesides, S. (1979). Collineations of projective pla of order 10, Parts I, II. J.Comb.Theory, (A)2 249-268; 269-277.

141] Wielandt, H. (1964). Finite Permutation Groups. New York: Academic Press.

142] Witt, E. (1938). Die 5-fach transitive Gruppen von Mathieu, Abh.Math.Seminar Hamburg, 12, 256-264.

143] Yamamoto, K. (1963). Decomposition fields of difference sets. Pac.J.Math., 13, 337-352.

INDEX